Natural Ecosystems

Natural Ecosystems

SECOND EDITION

W. B. Clapham, Jr.
Cleveland State University

Macmillan Publishing Co., Inc.
New York
Collier Macmillan Publishers
London

Macmillan Publishing Co., Inc.
866 Third Avenue, New York, New York 10022

Collier Macmillan Canada, Inc.

Library of Congress Cataloging in Publication Data

Clapham, Wentworth B.,
    Natural ecosystems.

    Includes bibliographies and index.
    1. Ecology   I. Title.
QH541.C45 1983   574.5   82-9898
ISBN 0-02-322520-3   AACR2

Printing: 1 2 3 4 5 6 7 8   Year: 3 4 5 6 7 8 9 0

ISBN   0-02-322520-3

Preface

A GREAT DEAL has happened to the field of ecology since *Natural Ecosystems* first appeared ten years ago. The role of natural ecosystems in a broad range of human affairs is much better recognized than it was at that time, but teaching about them has tended to return to departments of biology or geography, or to strong interdisciplinary programs in environmental studies. The computer revolution has brought about an increased understanding of the nature of systems and the usefulness of systems science in elucidating the principles of complex phenomena such as ecosystems.

The second edition of *Natural Ecosystems* reflects these changes. The book has been largely reorganized. Elementary systems science is used throughout as a tool for organizing and conceptualizing ecosystems. The chapters are smaller and more focused than in the first edition, and they are organized into parts that connect related issues. Not one paragraph in the entire book has escaped revision, with the sole exception of the excerpt from T. S. Eliot's *Dry Salvages* on page 1, which I personally find an eloquent statement of why students should be concerned with the principles of natural ecosystems.

The amount of terminology and jargon in this edition has been reduced. The book is oriented toward principles, and too many terms are more likely to obfuscate than to clarify, especially in an introductory textbook.

Despite these changes, the overall logic of the book remains the same as the first edition. It is intended as a first book that does not assume college-level background on the part of students using it. It casts its net broadly, including geological, geographic, and chemical aspects of ecosystems as well as biological. It begins with a brief overview of the components of ecosystems. Part II discusses the flow and cycle of energy and materials that sustain life. Part III looks at the populations and communities that constitute the living portion of ecosystems. Part IV describes several classes of ecosystems in some detail. The book concludes with an assessment of the implications of natural ecosystems in a modern society.

I would like to thank those who have aided in making this book possible. The manuscript was typed by Merilee Krick and Stacy Becker. I am grateful to the computer staff of the International Institute for Applied Systems Analysis in Laxenburg, Austria, for introducing me to the mysteries of word processing and to John Koenig of Computerland for keeping my own system operating (in its own arcane way), so that I could almost meet Macmillan's demands for a complete manuscript yesterday. The manuscript was reviewed by Peter W. Frank, Richard Alexander, Edmund Bedecarrax, Clyde

Hibbs, David Tilman, David Castillion, George Dalrymple, and Arthur Borror. Like all reviewers, their comments were often maddening, and they led to changes I had not expected to have to make. I didn't always accept the changes they suggested, but the book is greatly improved because of their suggestions. On the whole, their comments were the most complete and useful that it has ever been my honor to receive. Several reviewers went beyond the call of duty in their constructive comments and concern for the manuscript. The editor for this book was Gregory W. Payne of Macmillan. It's *de rigeur*, of course, for authors to thank their editors, but Greg Payne provided the finest support I have had from any editor I have ever been associated with on any manuscript. Not only did he assemble the panel of reviewers, he also sympathized, cajoled, and nudged with a remarkable degree of understanding and success. Finally, my wife and children deserve a special award for having put up with me during a difficult period.

W. B. C.

Contents

Natural
Ecosystems

Building
Blocks

T. S. ELIOT expressed the ambivalent relationship between modern society and the environment in which we live at the beginning of the third of his *Four Quartets*;

I do not know much about gods; but I think that the river
Is a strong brown god—sullen, untamed and intractable,
Patient to some degree, at first recognized as a frontier;
Useful, untrustworthy, as a conveyor of commerce;
Then only a problem confronting the builder of bridges.
The problem once solved, the brown god is almost forgotten
By the dwellers in cities—ever, however, implacable,
Keeping his seasons and rages, destroyer, reminder
Of what men choose to forget. Unhonored, unpropititated
By worshippers of the machine, but waiting, watching and waiting. . . .[1]

Eliot's brown god represented a natural order that preceded and transcended people both as a species and as a society. Modern technology is so powerful that we often forget the "brown god." Yet it is becoming increasingly clear that the natural order by which living organisms are bound to each other and to their environments is essential for the existence of any species on earth, including humans.

The complexity of the environment appears overwhelming when we first look at it and try to gauge its significance. The numbers of species of animals and plants are almost uncountable. The demonstrable interactions between them are even more so. Nonliving phenomena such as water, soil, and air sustain life and are regulated by it in ways that are not always clear. Nobody totally understands the patterns that link animals, plants, and the nonliving environment in any natural system. It was not until fairly recently that most scientists accepted that there were patterns, and there are still philosophical disputes over how to understand them.[2]

Ecology is the science of living systems. It deals with the interactions which link living organisms and their environments and determine the distribution and abundance of those organisms.[3] It is a growing science whose

1

focus is changing from the behavior of individual species to the patterns and laws governing the systems within which those species behave. It used to be defined as "the study of the distribution and abundance of organisms."[4] It concentrated on the organisms. The environment was simply the context for their behavior. It was "there."

Ecology is not strictly a biological science, although its primary focus will always be living things. We have learned enough to understand some of the patterns of the complex, self-sustaining natural systems of which living organisms are parts. We call these *ecosystems*. An ecosystem includes not only organisms, but also the nonliving components of the environment within which they are found. It also includes the bonds that link the living and nonliving components into a stable system.

To understand natural ecosystems requires a deliberately synthetic approach. However we approach them and divide them up into "pieces" for study, we must ultimately put the pieces back together and visualize the ecosystem as a unit. We can concentrate on their components or on the bonds that link them together. We can analyze them to figure out the laws by which they remain what they are, or we can study them as actual phenomena in real places. Their components are living and nonliving: plants, animals, soils, sediments, water, and air. Some of the bonds are concrete entities flowing among the components: energy and materials. Others are abstractions: the interactions among organisms that stem from their genetic heritage. Regardless of our perspective, the "grand design" that makes it all work is the ecosystem. There is an order to it, however obscured by the complexity of the real world. This order ensures the continuity of the ecosystem and thereby the persistence of life on earth.[5]

Notes

[1] From T. S. Eliot, "The Dry Salvages," from *Four Quartets* (New York: Harcourt Brace Jovanovich, Inc., 1941). Used with permission.

[2] Engelberg, J., and Boyarsky, L. L., 1979. The noncybernetic nature of ecosystems. *Am. Nat.*, **114**, 317–324; Patten, B. C. and Odum, E. P., 1981. The cybernetic nature of ecosystems. *Am. Nat.*, **118**, 886–895.

[3] Krebs, C. J., 1972. *Ecology, The Experimental Analysis of Distribution and Abundance*. New York: Harper & Row.

[4] Andrewartha, H. G., 1961. *Introduction to the Study of Animal Populations*. Chicago: University of Chicago Press.

[5] Patten, B. C. and Odum, E. P., 1981. The cybernetic nature of ecosystems. *Am. Nat.*, **118**, 886–895.

Components
of
Ecosystems

1

LIFE INHABITS a remarkably small proportion of the earth—only the surface layer of soil, oceans, lakes, streams, and that portion of the atmosphere inhabited by flying or floating organisms such as birds, bats, and bacteria. We call this zone the *biosphere*. Nowhere on earth is it thicker than a few kilometers, and there only in the deep oceans.

Ecology properly begins with life, because life is what distinguishes ecosystems from other natural phenomena. But it also involves the nonliving arena within which life proceeds. At its simplest, then, any ecosystem can be divided into two parts, living and nonliving, or *biotic* and *abiotic*. The abiotic portion includes the solid earth, liquid water, and the atmosphere. Each of these building blocks has properties of its own that determine its role in the ecosystem, and each interacts in certain ways with other components. Some of these properties vary in space or time and form the basis for differences between ecosystems, whereas others are essentially invariant throughout the biosphere. They may be physical, chemical, biological, or geological, and the patterns of the interactions between them may be simple or complex (Figure 1.1).

Ecosystems are *dynamic* entities: Their composition at any time reflects the complex of factors making them up. Changes in any factor, whether natural (as with seasons) or as a result of human perturbation, may have repercussions throughout the ecosystem as it responds first to the initial stimulus and then to the secondary changes induced by that stimulus. Understanding an ecosystem means understanding all of its phases, throughout the cycle of seasons and across the range of other variables that can influence it. It means using the tools and methodologies of many disciplines (including biology, geology, chemistry, and physics) to understand its components, and then synthesizing the observations and experiments into a single context, so that the unity and dynamic nature of the ecosystem become clear.

Water

Water is one of the most unusual natural compounds, and it is also one of the most important. Life began in the seas, and water in some form or other is essential to all life. Water is one of the main agents in weathering rocks and in the erosion and deposition of sediment. These are the key processes in soil formation and shaping landforms that provide the setting for all terrestrial ecosystems. Water covers some 71% of the earth's surface and is itself the

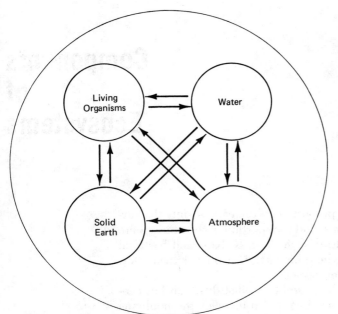

Figure 1.1
Schematic representation of an ecosystem.
It is characterized by the balances between
all its aspects, not by any one in particular.
Each of the entities shown is a reservoir for
energy, materials, and other things required
by ecosystems.

medium for several different ecosystems. The cycling of water between the hydrosphere and the atmosphere is a major factor in determining climate.

Many physical properties of water are unique. Under most conditions in the biosphere, it is a liquid that boils at 100°C and freezes at 0°C. It is both viscous and buoyant—much more so than air. Organisms can survive there without the specialized supportive structures needed by organisms that inhabit terrestrial environments. At the same time, mobile aquatic animals need to be streamlined if they are to move efficiently through the water, but they can swim using relatively simple movements.

The relationship between water's temperature and density (mass per unit volume) is equally unusual (Figure 1.2). When water is cooled from room temperature, it contracts and becomes denser as it gets colder, until it reaches a maximum density at 3.94°C. It expands as it cools below this point, unlike most other substances. Even more remarkable, it expands markedly upon freezing, so that ice at 0°C is about 9% less dense than water at the same temperature. Thus ice floats on top of a lake or stream, and it is unusual for aquatic ecosystems to freeze solid.

Water has an unusually high ability to absorb heat from its environment. It takes 1 calorie of heat energy to raise the temperature of 1 g of water 1°C. The amount of energy required to raise 1 g of other common substances 1°C is shown in Table 1.1. Water also exchanges a great deal of heat with its environment when it changes from a liquid to a gas or to a solid. Every gram of water at 100°C that changes to steam absorbs 540 calories of heat. It releases it when it changes back to water. Likewise, every gram of water at 0°C that changes to ice releases 79.7 calories, and reabsorbs it when it changes back to water. It takes over five times as much energy to vaporize a given amount of water at its boiling point without changing its temperature as it does to raise the water temperature from 0° to 100°C. It takes more heat to melt a given amount of ice at 0°C without changing its temperature

Figure 1.2
Relationship between temperature and volume of pure water: weight of 1 m.³ of water and density of water over a range of temperatures.

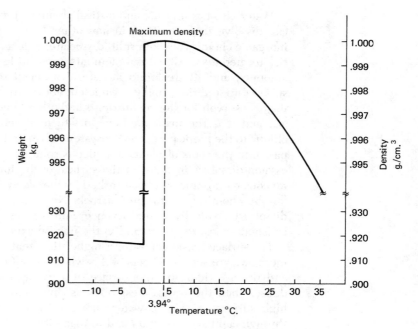

than to raise the temperature of the same amount of water from 0°C to over twice human body temperature (37°C.). No other common substance has these properties.

The consequences of water's thermal properties are many. Water can absorb or release much more heat than solid rock with much less change in its temperature. As a result, the temperature variation of an aquatic ecosystem is much less than that of a terrestrial ecosystem in the same geographic area. Also, as a lake is warmed by the sun, some of the water evaporates, absorbing heat that would otherwise have gone into raising the water temperature. Conversely, as a lake cools, ice formation releases heat into the environment and tends to slow down the rate at which further freezing takes place. Thus, water is a highly buffered medium: It is resistant to change. A fish living in a small pond in northern latitudes may have to withstand a total yearly temperature variation of about 25°C. This is much less than the temperature variation of some 65°C. that may be experienced by a nearby squirrel.

Water also influences climate on a regional and global scale. The moisture picked up by winds blowing over water also buffers the temperature of adjacent land areas. The climates of regions whose prevailing winds pass over oceans or large lakes are much less extreme than those that are not influenced by water. The global circulation patterns of the ocean (page 71) are a major mechanism for heat transfer from equatorial regions north and south toward the poles.

Table 1.1
Heat required to raise the temperature of 1 gram of several common substances 1°C.

From Smithsonian Institution, 1956.

	cal.
Fresh Water	1.00
Sea Water	0.93
Wet Mud	0.60
Moist Sandy Clay	0.33
Solid Rock	0.20
Copper	0.09

Water also has unusual and critical chemical properties. More substances can dissolve in water than in any other liquid. This is especially true of inorganic chemicals which split (dissociate) to form electrically charged entities termed *ions*. All natural elements are soluble in water, at least in trace amounts, and all are found in natural water at some place on the earth's surface. In addition, many organic chemicals are water soluble. Water is a major reservoir for storing nutrients and other biologically important materials, and it is the main medium in which these materials move from the abiotic to the biotic part of an ecosystem. Even in the driest desert, nutrients pass into the roots of plants in aqueous solution; when animals breathe air, oxygen dissolves in water at the surface of the lung before it can cross the mucous membrane and be absorbed by the blood. Water is also the medium for the chemical weathering of rock. No rock is so resistant that it is not dissolved slowly by water. Even granite breaks down into soil under the constant exposure to rain and to the normal exudates of plants and animals.

The surface tension of water is higher than that of any natural liquid except mercury. We see the effects of this when we notice things like pollen, dust, and water striders on the surface of a water body even though they are denser than the water. Less obvious, but more significant, however, the high surface tension of water allows soils to retain a significant amount through capillary attraction and to make it available to terrestrial plants. In addition, surface tension influences several basic physiological and biochemical reactions within living organisms.

The Solid Earth

It is easy to take the solid earth for granted: it is there, and it is so variable that it is easy to ignore the variations. But the earth is critical to the operation of ecosystems. It is useful to divide the solid portion of the biosphere into three basic aspects: rocks, sediments, and soils.

Rocks are consolidated units of the earth's crust made of minerals that have come together by hardening of sediments (sedimentary rock), by solidification from a molten mass (igneous rock), or by alteration of a preexisting rock at high pressure and temperature (metamorphic rock). Rocks are the parent materials from which sediments and soils are derived. The combination of rocks, relief, and climate determines the patterns of erosion and the subsequent development of landforms. In addition, porous rocks and sediments are the media for the storage and movement of groundwater.

The properties of rock have a tremendous influence on the characteristics of sediments and soils derived from them. All rocks exposed at the surface undergo *weathering*. Weathering includes physical processes in which large rock fragments are broken up by thermal expansion and contraction or gravity, as well as chemical processes involving solution of certain minerals by water, chemical changes from one mineral to another, and biochemical reactions involving both water and acid exudates from bacteria, plants, and animals. The patterns of rock weathering result from the interactions of many variables, including the chemistry of the minerals in the rock, the amount of water available as a medium for chemical weathering, the materials dissolved in the water, and the types of organisms in the area. In return, the

sediments and soils resulting from this breakdown influence the water relations of the ecosystem and the types of organisms found in it. Living things
and the solid earth each exert a strong influence on the other, and the dynamic balance between them allows the ecosystem as a whole to change as a
unit.

The distribution of rocks at the earth's surface influences the distribution
of ecosystems in several ways. Landforms such as mountain ranges or plains
have local effects on the structure of ecosystems, and they also influence
climate on a continental scale. Many of the chemical properties of rocks are
passed on to sediments and soil derived from them.

In an area undergoing erosion, soft, easily weathered rocks tend to be
selectively removed and form depressions in the landscape. Less readily
weathered rocks tend to remain as high points. Regional topography depends on the vertical distance between the high and low points and on the
steepness of the gradients connecting them. The factors determining the
shape of an erosional landscape include the regional climate and the relative
weatherability of the exposed rocks under the climatic conditions found in
the area.

Landforms influence ecosystems in different ways. Organisms adapt differently to different topographic conditions such as cliffs or rolling country-
side. On a larger scale, the distribution of landforms can affect climatic patterns. In the middle latitudes, the dominant direction of air circulation is
from west to east. Moisture-laden air moving east off the oceans rises as it
meets mountains oriented in a north-south direction, such as the Sierra Nevada of North America or the Andes of South America. But it cools as it rises,
reducing its ability to carry moisture. The result is heavy rainfall on the
western slopes of the mountains and a "rain shadow" with very little rainfall
to the east. Thus in North America and Argentina, modern deserts and
semiarid grasslands exist east of the mountains. The fossil record indicates
that the deserts and grasslands in North America did not exist prior to the
rise of the Rocky Mountains. Indeed many of the main plant groups that now
characterize desert areas evolved as a response to the climate change which
accompanied the changing landforms, and the adaptation of many animals
evolved to follow the changing flora. In central Europe, on the other hand,
where the major mountain ranges tend to run more in an east-west direction, deserts have not developed and the climate is more uniform.

Rocks are also the medium for groundwater. Many rocks are quite porous,
and fluids such as water (or oil or gas) can exist in the pore space between
rock particles. The fluids can move through the rock if the pore spaces are
interconnected. Thus water entering a rock in one location may actually flow
through the rock to another location where it can be removed, either in a
spring or by an artesian well. Such a rock is called an *aquifer*. These may
extend over vast areas, and they are important sources of water in many
areas.

Sediments are broken-up rock fragments that may or may not be chemically altered by weathering. In many ways, sediments are intermediate between rocks and soils, overlapping in certain characteristics with each. Just
as rock particles are worn away by wind and water, sediments are deposited
when the erosive-transportive force is no longer sufficiently powerful to
erode or transport materials. Depositional areas can be any place where the
velocity of the transportive medium decreases, such as flood plains along

rivers, in lakes, or in the sea. The oceans are the ultimate depositional site for all sediments.

Sediments interact most strongly with living organisms in aquatic ecosystems. They are the bottom of lakes and oceans, as well as of streams that do not flow directly over bedrock. They comprise not only mineral matter eroded from adjacent land areas but also the partially decayed remnants of animals and plants that lived in the ecosystem. Nutrients collect in the sediments, and many aquatic ecosystems depend to a large extent on the cycling of nutrients back and forth between the sediment and the water. Sediments are also the rooting medium for aquatic plants and home for burrowing animals.

Soils are much more complex than simple sediments, although they are derived from sediments. They comprise a complex mixture of rock fragments, highly altered minerals, organic debris, and living organisms. They are characteristic only of terrestrial ecosystems. Soil formation depends on the characteristics of the parent rock or sediment, the climate (especially temperature and rainfall), the types and numbers of organisms living in the area, the topography, and time. Soils are the source of almost all nutrients and much of the water available to organisms in terrestrial ecosystems. They are the structural foundation for rooted plants, and they support the entire weight of all organisms living in a terrestrial ecosystem.

The Atmosphere

The atmosphere is an ocean of air which blends into outer space some 1000 km or so above the earth's surface. We can distinguish four concentric layers on the basis of temperature (Figure 1.3). The lowest layer, or *troposphere*, is roughly 10 km thick (it is somewhat thicker in the equatorial region than it is at the poles). Its temperature decreases steadily from the earth's surface to the top. It is a mixture of several gases, most of which are fairly constant in their abundance (Table 1.2).

These gases are fairly inert, with the exception of oxygen, carbon dioxide, methane, and ozone. The first two are essential for life: Carbon dioxide is taken up by photosynthesis in green plants, and oxygen is given off; the reverse occurs in respiration. Also, combustion of fossil fuels removes oxygen and adds carbon dioxide to the atmosphere. Because oxygen is so abundant in the atmosphere, the variations induced by biological activity and burning fossil fuels do not show up. But carbon dioxide is so rare that even small changes in the oxygen/carbon dioxide balance can bring about noticeable changes in the CO_2 concentration. Methane also varies from place to place: It is released into the atmosphere by petroleum and natural gas production, and it is given off by living organisms in certain areas. Ozone is formed in the atmosphere by photochemical reactions of oxygen molecules, and its abundance in the atmosphere depends on the balance of factors that control the ozone-creating and ozone-destroying reactions. These vary by season, time, and place.

Water vapor and dust are also present in the troposphere, but their concentrations are even more variable than the methane, ozone, and carbon dioxide. The concentration of water vapor ranges from virtually 0% to more than 4%. Its abundance depends most on altitude and temperature. It is

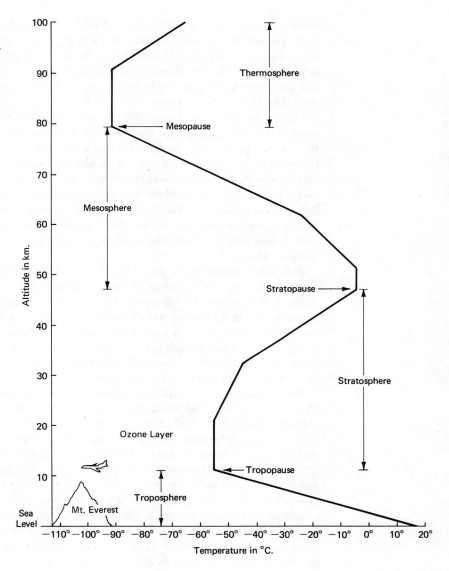

Figure 1.3
Temperature stratification of the atmosphere. [Redrawn, with permission, from U.S. Air Force Cambridge Research Laboratories, *Handbook of Geophysics and Space Environments.* Copyright © 1965, by the United States Air Force.]

Table 1.2
Composition of the Dry Atmosphere by Volume

Reprinted, with permission, from U.S. Air Force Cambridge Research Laboratories, *Handbook of Geophysics and Space Environments.* Copyright © 1965: United States Air Force.

Constituent	Percentage
Nitrogen (N₂)	78.084
Oxygen (O₂)	20.9476
Argon (Ar)	0.934
Carbon Dioxide (CO₂)	0.0314°
Neon (Ne)	0.001818
Helium (He)	0.000524
Methane (CH₄)	0.0002°
Krypton (Kr)	0.000114
Nitrous Oxide (N₂O)	0.00005
Hydrogen (H₂)	0.00005
Xenon (Xe)	0.0000087
Ozone (O₃)	0.000001°

*These components are highly variable.

most common very near the ground and almost entirely absent above 8 to 10 km, and it is more abundant in warm air than in cold air. Dust, on the other hand, is even more limited to the lower levels of the atmosphere than is water vapor. Its abundance is related to wind velocity, volcanic activity, and any other factor (such as human activity) that makes it easier for wind to pick up soil from the surface.

The troposphere has by far the greatest influence on ecosystems of any level of the atmosphere. It is the medium of our weather, and global circulation of air masses and weather systems is an important mechanism for distributing heat around the globe. It is also the source of the gases used directly by living organisms in their metabolism. Just as it provides the oxygen and carbon dioxide needed for life, its composition is strongly influenced by living things. Indeed, the Earth's original atmosphere was very different from the present atmosphere and that the large proportion of oxygen we now find is due largely to the release of oxygen gas from archaic plants through photosynthesis.

But the troposphere is not the only stratum influencing ecosystems in significant ways. The *stratosphere* is the air mass extending from the top of the troposphere to a level about 50 km above the surface of the earth. As one ascends, the temperature rises from about $-55°C$ to about $+5°C$. At the top of the stratosphere, a second temperature reversal takes place, and the temperature begins to drop, reaching a minimum about $-95°C$ at a level some 80 to 90 km above the earth's surface. Above this level, the temperature increases steadily into outer space.

The stratosphere is strikingly different from the troposphere in several significant aspects. Water vapor is virtually absent. The only clouds are very thin wispy clouds formed of tiny ice crystals. Ozone (O_3), formed from oxygen by a *photochemical* reaction in which solar energy splits apart oxygen molecules (O_2) to form atomic oxygen (O), is present there in significant quantities . Oxygen atoms are very unstable and reactive. They combine readily with molecular oxygen to form ozone. Both of these reactions are reversible: They can proceed either so that molecular and atomic oxygen combine to form ozone or so that ozone breaks down to form molecular oxygen. On the whole, the ozone concentration in the stratosphere is constant. This indicates that it is in dynamic equilibrium and that it is being produced from oxygen as fast as it is broken down.

Ozone absorbs ultraviolet radiation from the sun. In fact, the reason the stratosphere becomes warmer with increasing altitude is that the ultraviolet energy absorbed by the ozone is transformed into heat.[1] This is vitally important for ecosystems, since ultraviolet radiation is lethal to most living organisms, and its absorption in the ozone layer prevents it from reaching the surface of the earth. Above the stratosphere, neither the composition of the atmosphere nor its importance to natural ecosystems is well understood.

Living Organisms

Living organisms have a more complex role in ecosystems than the abiotic phases because they have an adaptive dimension that has no analogue in the abiotic parts of ecosystems. The structure, behavior, and metabolism of any organism result from its genetic makeup. These adaptations determine the way an individual gets its food, interacts with others of its own kind, and interacts with individuals of other species.

One can almost view the interactions among species as a vast drama played out in a natural global theater. But this is not a drama. There is no playwright and no artistic purpose is served. It is anthropomorphic to speak of ecosystems as having a "goal" or a "function." Discussions of the goals and functions of natural ecosystems are theological, not scientific. But it is meaningful to note the unity and direction in the dynamics of ecosystems. Survival of individual species means more than the preservation of genetic material. It also means preserving the role of the species in the ecosystem. Nature is complex, but it is not chaotic. Organisms have specific resource needs. If ecosystems were disorganized and chaotic, living things could not depend on them to meet their needs, and life as we know it could not last very long. Living things have adaptations that give them a role in an ecosystem, and the ecosystem provides an order that allows life to persist.[2]

One cannot speak of the adaptations of a rock. It may break down to yield a soil, but natural selection does not cause it to be replaced by some other rock that will yield a different kind of soil better adapted to the ecosystem. In the same way, climate depends on the interactions of several aspects of the ecosystem, and it may change as these factors change. But it is not adapted to the ecosystem in any way. The materials found in the water in a pond or stream or in the air we all breathe may result from interactions with living organisms, but they are not adapted in any sense to any ecosystem. Only living organisms have genetically determined adaptations to an environment, and only living organisms have resource requirements that must be met if they are to survive.

Natural Selection and Evolution

Evolution describes the process by which species come to possess adaptations. Its mechanism is *natural selection*. There is nothing mysterious about either. But it is useful first to distinguish three levels for understanding living organisms: the individual, the population, and the community.

Individual organisms live and die. Their adaptations include all the physical and biochemical mechanisms the species uses to acquire resources from the environment, to transform these resources into useful products, to excrete waste materials, and to fulfill the other necessities of life, such as growth, movement, and reproduction. Other adaptations may improve the survival of the organism in other ways, such as by conferring resistance to physical stress through various biochemical or behavioral adaptations or by protecting the organism from potential predators through protective coloration, distastefulness to the predators, or active defense mechanisms. Individuals also possess genes and pass them from one generation to the next when they reproduce. This is the physical mechanism for passing on hereditary characteristics.

The *population* comprises all the individuals of a given species in an ecosystem. But its significance is more than that of a number of individuals. An individual competes with others in the same population for the environmental resources they all need. But not all individuals are identical. Populations contain variation: Even fundamental adaptations such as hair color or size may differ slightly from individual to individual, and not all members of the population are equally likely to survive and reproduce under the dynamic stresses of competition for these resources.

The *community* includes all the populations in an ecosystem. Its structure involves many types of interactions among species. Some of these involve the acquisition and use of food, space, or other environmental resources. Others involve nutrient cycling through all members of the community and mutual regulation of population sizes. In all of these cases, the structured interactions of populations lead to situations in which individuals live or die.

Evolutionary change is a change in gene frequency. The patterns of variation in a population change from one generation to the next. Individual organisms interact with each other within the context of their overall environment. Some survive; others do not. Those that do, and produce the most offspring, contribute most to the genetic basis of their population's posterity. Those that do not reproduce contribute nothing. Natural selection acts through individuals, but it influences the characteristics of the population. Differential reproduction is the critical factor controlling the frequency of different genes surviving in the population.

The best known example of natural selection operating in modern populations is pesticide resistance in crop pests. Prior to the widespread use of modern pesticides that began in the mid-1940s, crop pest populations contained a certain amount of genetic variability for resistance to DDT and similar pesticides. Nobody can be sure how much, because natural selection can act only with respect to a concrete environmental factor. Natural selection in the absence of pesticides can not lead to change in the frequencies of genes conferring resistance to chemical pesticides. However, once spraying started, individuals that happened to possess resistant genes survived much more frequently than those that did not, and they were able to pass a disproportionate percentage of their genes on to the next generation. As a result, the introduction of pesticides into the environment provided a tremendous selective pressure to increase the frequency of resistant genes in the pest populations. This is not an abstraction, however. This increase in gene frequency is equivalent to an observable rise in pesticide resistance in the population. It has been observed in practically every pest that has ever been the target of practically every pesticide that has ever been tried.[3]

In effect, using pesticides represents a change in the abiotic conditions of the environment. There is no basic difference in the selective pressure from this source and from climatic or other fundamental change. Changes in the gene frequency of pesticide resistance is not inherently different from changes in the frequency of other genes. As we shall see in Chapters 5 and 6, any sort of change in an ecosystem can influence many populations in addition to the primary target. Natural selection provides a powerful mechanism for the development of ecosystems.

Feedback in Ecosystems

Most interactions among living organisms, or between organisms and the abiotic phases of the ecosystem, are mutual: The activities of one population are met by responses from the other. These often involve *feedback*, the most important controlling principle in complex systems of this sort. Feedback occurs when an organism's behavior depends not only on some original stimulus but also on the results of previous behavior. How this works can be shown schematically. Figure 1.4a shows behavior in the absence of feed-

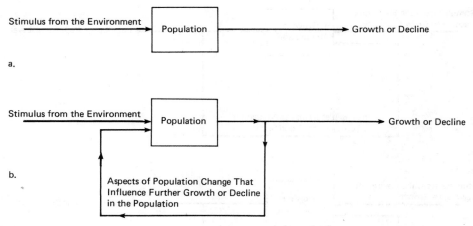

Figure 1.4
Population growth with and without feedback. a. without feedback. b. with feedback.

back: The stimulus is independent of the population, and the result is prede-
termined. As an example, the seeds of many desert flowers remain dormant
in the soil—often for years—until wetted by a heavy rain. They germinate
within hours, grow quickly, bloom, and die. The rain is the stimulus, growth
and blooming are the responses. Figure 1.4*b* shows a feedback system.
There may be some initial stimulus external to the organism, but its behavior
in the future depends to a large extent on the results of its past behavior. For
example, a cheetah beginning to chase an antelope will run straight toward
it. But the antelope responds by running away. The cheetah must compensate for this and change its direction if it expects to eat; that is, it must adjust
its behavior for the change in the system (i.e. the antelope's running away)
induced by its own behavior.

There are two types of feedback, positive and negative. *Positive feedback*
leads to self-sustaining change; *negative feedback* maintains a dynamic homeostasis, or steady state. Figure 1.5 diagrams each type, using population
growth as the example. In the first case, a small population comprising a
dozen individuals reaches an island in the middle of the ocean. Conditions
are favorable, and they reproduce. The first generation is small, perhaps 25.
But the second generation doubles once again to 50. As long as the number
of offspring depends mainly on the size of the breeding population, the
population will continue to double each generation. The tenth generation
will be over 10,000. Of course no population can continue to increase indefinitely at this rate, but this kind of positive feedback is typical when breeding
populations get established in new areas. Examples include rabbits in Australia and gypsy moths in Massachusetts.

In the second case, the population is controlled by its food supply. There
is an equilibrium population density that can be maintained by the available
level of food. If, for some reason, the population increases above this level,
there is insufficient food to go around, and more individuals than usual die
from starvation. Conversely, there is more than enough food to go around if
the population falls below the equilibrium level; the animals are well nourished, and die-off is smaller than normal. In either case, the population
density tends to return toward the equilibrium level. Examples of this sort
are well known; several will be described in detail in chapter 5.

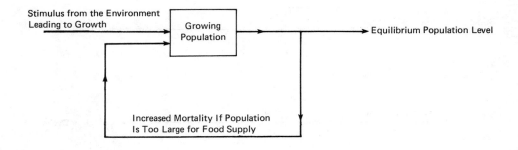

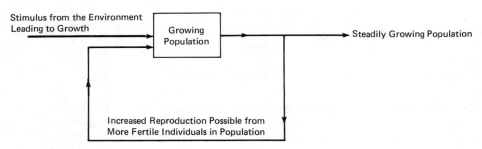

Figure 1.5
Population growth in systems with positive and negative feedback. a. Negative feedback, leading to self-regulated homeostasis. b. Positive feedback, leading to self-sustained further change.

Feedback-based relationships are very common in nature. All populations in any real ecosystem participate in several different feedback loops at once. Some of these involve negative feedback, while others are positive.

Quantifying Living Organisms

We need to be able to deal quantitatively with living organisms in order to understand their interactions. Sometimes it is sufficient to count the number of individuals in the population, or the number of individuals in each of the several age groups within the population. Sometimes it makes more sense to weigh organisms than it does to count them; this is especially true if species of different size are being compared. The total weight of living material in any given species in an ecosystem is termed the *biomass*. Biomass can be measured in terms of live weight, but the water content of different species varies so much (e.g., a jellyfish is about 99.9% water by weight, whereas people are only 71% water) that it is more common to dehydrate the organisms and measure dry weight.

Time

Time is a fundamental variable in everything that happens in an ecosystem. Understanding natural ecosystems requires that we view their development as functions of time as well as of other factors. We can distinguish at least four different perspectives of time which grade into one another but are all meaningful in ecology: very short time, human time, historical time, and geological time (Figure 1.6).

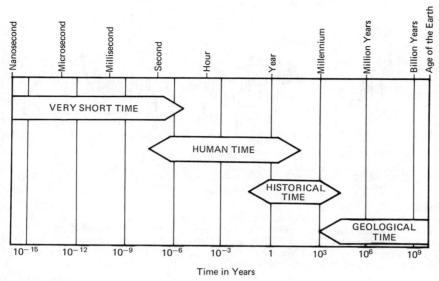

Figure 1.6
Diagrammatic representation of four different perspectives on the time continuum.

Very short time is the scale on which "instantaneous" changes occur. Some features of the ecosystem must be viewed on this time scale because they simply happen too fast to be measured in any other way. These include such phenomena as electrical discharges in lightning, movement of nervous impulses in an animal, and numerous chemical reactions that occur in both the living and nonliving parts of the ecosystem.

Human time is the span of time in which most of us think. "Rapid" changes can be measured in seconds or minutes; these include actions such as an owl seizing a rabbit or a mass of rock moving in a landslide. "Slow" changes are measured in days, months, or years; these include passing of seasons or the growth of an animal or plant. Most observations of ecosystems use this time scale. It is also the common time scale in most human affairs, and "long-term planning" seldom means more than ten to twenty years ahead.

Historical time refers to the span of time for which written records are necessary and where trends extend over decades or millenia—intervals too long to be studied adequately by one person. Ecosystems seen in the perspective of historical time look very different from their appearance in human time. For instance, Figure 1.7 shows the average annual temperature of the Eastern seaboard of the United States centered about Philadelphia between the mid-18th century and the present. The year-to-year fluctuations are substantial, and a resident of Philadelphia would be very conscious of them. But the data viewed over historical time indicate a stable average annual temperature about $12\frac{1}{2}$°C.

Some characteristics of ecosystems must be examined from the perspective of historical time because they are naturally very slow. These include processes such as soil formation and some kinds of landform alteration. Many fluctuations that seem striking in human time (such as the temperature of Philadelphia or populations of fish species in the Pacific Ocean) even out in historical time. They are generated by random influences on the ecosystem and may obscure significant underlying characteristics. Historical records of

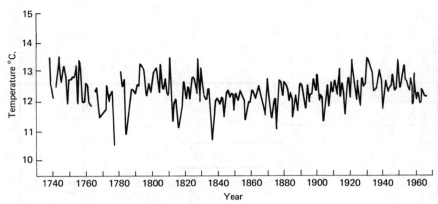

Figure 1.7
Annual temperatures for the eastern seaboard of the United States for the period 1738 to 1967; a representative reconstructed synthetic series centered on Philadelphia. [Reprinted, with permission, from H.E. Landsberg, "Man-Made Climatic Changes," *Science* 170 (December 18, 1970), 1265–1274. Copyright © 1970 by the American Association for the Advancement of Science, Washington.]

ecosystems can indicate what it "ought to be" or how it has changed under human influence. For example, the fossil records of the inhabitants of lakes show the types of organisms that lived there before the coming of civilization. This has been suggested several times as the only means of determining the "natural" inhabitants of a lake, either because changes since the rise of pollution levels have been so great that scientists simply do not know who the original inhabitants were, or because observations of present communities show such great fluctuations that it is impossible to distinguish "forests" from "trees."[4]

Geological time refers to the very long-term changes within ecosystems. Just as historical time provides a benchmark for understanding the present, seeing ecosystems in geological time shows that they are not and have never been static. For instance, if we followed the temperature of Philadelphia backwards through geological time, it would not remain constant at $12\frac{1}{2}°C$. It would drop below freezing during the advances of the glaciers in the Ice Age, and then rise to a high point some 100 million years ago when dinosaurs swam in the primordial Delaware River and subtropical vegetation extended into Canada. The very slow changes in ecosystems that can be seen only in geological time are as significant as those that are measured in historical or human time; these include broad climatic changes, gross modifications of landforms, and evolution of organisms.

The present is a period of rapid change, much of it the result of human intervention in the environment. People are now the most powerful agents of natural selection on the earth, especially in the simplified ecosystems of fields and cities, through our use and misuse of agricultural chemicals, and our production of urban wastes. Many changes that would normally take place over geologic time have been compressed into human time. We have been selective and have altered the rates of only certain changes. But natural ecosystems are balanced systems with an astounding array of checks and balances that allow elements within the ecosystem to respond and adapt to changes in other elements. When people increase the rate of change beyond what the ecosystem can respond to, we upset the natural balances and turn a stable system into an unstable one.

A perusal of the record of ecosystems in the geologic past shows that this is not the first time unstable systems have existed (although it is the first time that the destabilization has resulted from the activities of thinking creatures). Several periods in earth history have seen major changes in ecosystem structure. In every case, the unstable systems of the past have regained an equilibrium, but only after an interval of great change, generally accompanied by massive extinctions of major life forms and extensive alteration in the structure of contemporary biological communities.

Circumscription of Ecosystems

Ecology is a practical science; it concerns how real systems operate, sometimes systems with considerable economic significance. But it is a systems science that deals explicitly with the interconnections among elements in the system. One of the watchwords of environmentalism has, in fact, become "Everything is connected to everything else." This is true in some very important ways. But there is a danger in taking it too seriously, since it suggests that one must study "everything" if one is to know anything. It is obviously impossible for even a large research team to study everything.

Practical ecological investigation depends on our being able to identify and delimit an ecosystem as a meaningful object for study. This would be simple if an ecosystem were closed: that is, if it were surrounded by an invisible box with nothing entering or leaving. The boundaries of the system could be defined as the walls of the box. But ecosystems are virtually never closed. Almost all the energy that allows living organisms to exist on earth comes ultimately from the sun; the exceptions are heat issuing from fissures in the deep ocean trenches and certain energy-rich chemicals. Sediment is eroded from mountains, transported by rivers, and deposited in lakes or oceans. Animals migrate from one area to another: The arctic tern summers in the Arctic and winters in southern South America, and the salmon spawns in freshwater lakes and rivers, but spends its adult life in the open ocean.

However, the natural intercourse between ecosystems does not preclude our defining meaningful boundaries. Natural boundaries such as shorelines clearly separate aquatic ecosystems from terrestrial ecosystems. Some elements of the system may cross the boundary; others do not. Other boundaries are less clear, such as that between a river and a lake, or between a rapids area in a river and a smoothly flowing area of the same river. The choice of limits for an ecosystem of study is a pragmatic one: boundaries must make sense in the light of the study being made, and the scientist must be sensitive to the features of the system that are not limited by those boundaries. For instance, the fact that different organisms live in a riffle area and in a smoothly flowing pool area of the same stream may be reason enough to consider the riffle an ecosystem even though the water and any dissolved or suspended materials it may contain flow through it unimpeded. Likewise, it may be advantageous to consider an entire forest an ecosystem if one is studying forests. But it may make equal sense to consider the leaves of certain species of trees an ecosystem if one is studying certain insects and if a balanced community exists within those leaves.

It is often useful to study a model ecosystem in the laboratory by establishing a controlled environment that removes some of the variables found in the field. This is often the only way of simplifying the ecosystem to a point where some of the more complex interactions make sense. In this case, the model ecosystem consists of the culture chamber. However, the scientist must do a thorough job of defining the ecosystem in the field before establishing a meaningful microcosm. This is the only way to get a clear idea of the things that cross the boundary of the system. The purpose of a microcosm is to mimic the real world in a significant way. The boundaries of the culture chamber are as arbitrary or as meaningful as the boundaries of the system in the field upon which the microcosm is modeled. Ecosystems have inputs from other ecosystems and outputs to other ecosystems. These inputs and outputs are commonly critical to the behavior of the ecosystem and must be included in the model.

At best, the study of ecosystems is an exceedingly complex task. Regardless of the size of the study area or its relationship to other areas, certain things remain constant. An ecosystem is a highly integrated series of interactions between the nonliving earth and living organisms. We cannot understand the workings of a natural ecosystem by looking only at the living portion or only at one of the abiotic phases. We cannot understand them even with a broad systems analysis if we ignore or understate the inputs from and outputs to other ecosystems. Ecosystems are complex dynamic entities. There is no shortcut to understanding either their complexity or their dynamism, but herein lies their excitement.

Notes

[1] Craig, R. A., 1968. *The Edge of Space: Exploring the Upper Atmosphere*. Garden City, N.Y.: Doubleday & Company, Inc.

[2] Patten, B. C. and Odum, E. P., 1981. The cybernetic nature of ecosystems. *Am. Nat.*, **118**, 886–895.

[3] Brown, A. W. A., 1978. *The Ecology of Pesticides*. New York: John Wiley & Sons.

[4] Edmondson, W. T., 1969. Eutrophication in North America, in *Eutrophication: Causes, Consequences, Correctives*. Washington: National Academy of Sciences, 124–149.

Further Reading

Ashby, W. R., 1956. *An Introduction to Cybernetics*. New York: John Wiley & Sons, Inc.

Beckner, M., 1959. *The Biological Way of Thought*. Reprinted 1968 by The University of California Press.

Hardin, G., 1963. The cybernetics of competition: A biologist's view of society. *Perspectives in Biology and Medicine*, **7**, 58–84.

Hutchinson, G. E., 1979. *The Kindly Fruits of the Earth: Recollections of an Embryo Ecologist*. New Haven, Conn.: Yale University Press.

Landsberg, H. E., 1970. Man-made climatic changes. *Science*, **170**, 1265–1274.

Likens, G. E., and Bormann, F. W., 1974. Linkages between terrestrial and aquatic ecosystems. *BioScience*, **24**, 447–456.

Murphy, R. E., 1968. Landforms of the world. *Ann. Assn. Am. Geog.* Map Supplement 9.

Rumney, G. R., 1970. *The Geosystem: Dynamic Integration of Land, Sea, and Air*. Dubuque, Ia.: Wm. C. Brown Company, Publishers.

Smithsonian Institution, 1956. *Smithsonian Physical Tables*, Smithsonian Misc. Coll. Vol. **120**.

Trewartha, G. T., Robinson, A. H., and Hammond, E. H., 1967. *Elements of Geography*, 5th ed. New York: McGraw-Hill Book Company, Inc.

Trewartha, G. T., 1968. *An Introduction to Climate*, 4th ed. New York: McGraw-Hill Book Company, Inc.

U.S. Air Force Cambridge Research Laboratories, 1965. *Handbook of Geophysics and Space Environments*. Bradford, Mass.: U.S. Air Force Cambridge Research Laboratories.

Flow and Cycle

II

SYSTEMS ARE HELD together by the interactions among their elements. In a typical human system (such as a corporation or a university), these interactions are in the form of information designed to allow the system to meet its specified purpose. "Information" in this case comprises orders, directives, production or purchasing data, and so on.

Ecosystems are held together by the interactions among their elements in much the same way, except that they are not systems organized by human leaders and directed toward particular goals. They are self-organizing systems whose internal structure allows them to maintain themselves.[1] In addition, many of the most significant feedback signals are not information but rather concrete phenomena.

Information and concrete phenomena have very different roles in feedback-regulated systems. Information is an abstract representation of a phenomenon whose significance comes from its interpretation. For example, a hawk circling over a field sees a rabbit. Visual information about the rabbit's presence and position allows the hawk to begin its attack. The rabbit's responses are channeled through the hawk's eye and interpreted continuously by its brain to allow it to adjust its attack. Concrete phenomena, on the other hand, require actual channels for moving the phenomena through the system, storage devices, and mechanisms for obtaining needed quantities and eliminating excesses. Energy, nutrients, metabolic wastes, and similar materials move from population to population. The populations' survival depends on their meeting their needs for these materials, and the feedback signals that regulate the ecosystem are actually material phenomena.

Many information-carrying channels exist. They include electromagnetic radiation (light), as in the visual image of the rabbit interpreted by the hawk's brain. Signals may be chemical: Hormones secreted by the brain are interpreted by an organ of the body and translated into action by the organ; urine deposited in key places by mountain lions to mark their territories is recognized by other mountain lions and is the basis of the geographic distribution of individuals within the population. Signals may be physical: An accounting report on a piece of paper serves as the basis for planning next year's production in a factory; bees returning to the hive after finding a source of nectar do a dance that indicates explicitly the direction and the

21

distance from the hive to the flowers. Signals may be aural: Speech is one of the most efficient ways of transmitting information from one person to another; many animals use sounds to challenge others of their species to mating battles, to identify their young, or to terrify a potential prey (or to bluff a potential predator).

Concrete phenomena are also of several sorts. Energy reaches the biosphere as light. It is transformed and stored in organic chemicals that can be utilized by the body to power its metabolism or passed to another organism as food. Nutrients are particular elements that become bound up in these organic chemicals and give them the particular properties that make them useful in the body. Some organisms secrete chemicals that are toxic to animals or to plants. In each case, the energy or material in question affects organisms and brings about responses, just as receiving information affects organisms and brings about responses. But the mechanism by which the concrete phenomena act shows much more of a "brute force" approach.

It is easy to see why human systems tend to be organized around information networks: They are much less cumbersome, and sending a "cheap" (or low-energy) representational signal that can be interpreted by the receiver is much easier and faster than transmitting a concrete phenomenon that does not need to be interpreted. Nevertheless, the primary network of signals linking the elements of ecosystems is concrete. This network operates much more slowly and less efficiently than a modern electronic device, but ecosystems based on this primary network of concrete signals have been preserving the order needed to sustain life for 2 to 3 billion years.

The key concrete signals are the energy and materials required in biological tissues. Energy is transitory in an ecosystem: It enters as light, is transformed into biochemical energy, and is either degraded to heat or stored in sediments. It flows once through the system. Materials, on the other hand, are based on chemical elements that are neither created nor destroyed. They may be involved in many different kinds of organic or inorganic molecules, they may react with many different things, and they may move from one place to another. But they are not degraded, and they are not lost. They cycle interminably through the system.

Ecosystems are dynamic entities. How they look at any one time is a function of how all of the populations and the abiotic environment are linked, mainly through the primary network of concrete signals. It is tempting to assume that ecosystems are in dynamic equilibrium: that is they do not change, on the average, because the import and outflow of any parameter balance one another. Sometimes this is indeed the case. Often, however, the materials are not in balance; there may be a net inflow or net outflow of something of critical importance to the ecosystem. It is often useful to speak of budgets, which are detailed accountings of the concrete signals that show the gross input, the gross outflow, and the net surplus or deficit for each type of signal. Most ecosystems have surpluses in some things and deficits in others. One of the most fundamental traits of ecosystems is the way in which these imbalances are overcome.

Note

[1] Patten, B. C. and Odum, E. P., 1981. The cybernetic nature of ecosystems. *Am. Nat.*, **118**, 886–895.

Energy in
Natural
Ecosystems

2

ENERGY CAN BE defined as *the capacity to do work*, whether that work be on a gross scale, such as raising mountains and moving air masses over continents, or on a small scale, such as transmitting a nerve impulse from one cell to another. Three sources of energy account for virtually all the work of the ecosystem: gravitation, internal forces within the earth, and solar radiation. The last is of greatest interest to us. It is important not so much because there is more solar energy than any other kind but rather because the sun is the ultimate source of energy for virtually all living things. In addition, solar energy heats the earth so that ecosystems can exist, and it drives many other natural phenomena, from atmospheric circulation to cycling of water through the ecosystem.

Solar Radiation Budget of the Earth

The sun produces immense amounts of energy through thermonuclear fusion: Small atoms such as hydrogen fuse, much as in a hydrogen bomb, to form larger atoms such as helium, releasing energy in the form of electromagnetic waves. Electromagnetic radiation includes a broad spectrum, from X-rays to radio waves (Figure 2.1). However, almost all the sun's energy is in the ultraviolet, visible, and infrared radiation bands; indeed about half of the total radiation of the sun is visible light. The intensity of radiation at any distance from the sun is termed the *solar flux*. It can be measured in cal/cm², or *langleys* (ly). Because the distance from the sun to the earth is fairly uniform year-round, the amount of energy reaching the outer atmosphere is within 5% of a constant quantity of energy (1.94 cal/cm²/min) termed the *solar constant*.

The biosphere depends on the amount of energy that reaches the ground, or *insolation*, not the amount that reaches the outer atmosphere. And insolation varies widely from place to place. The atmosphere alters the solar radiation spectrum significantly as light passes through it to the surface of the earth. The changes in quantity and spectral quality of solar radiation as it passes through the atmosphere depends on the time of day, the season, the amount of cloud cover, and other factors. The net result of these variables is a seasonal and latitudinal distribution of insolation, as shown in Figure 2.2.

We can also be more specific and describe the overall disposition of solar energy between the outer atmosphere and the ground. Figure 2.3 summarizes the disposition of light energy at midday under average summer condi-

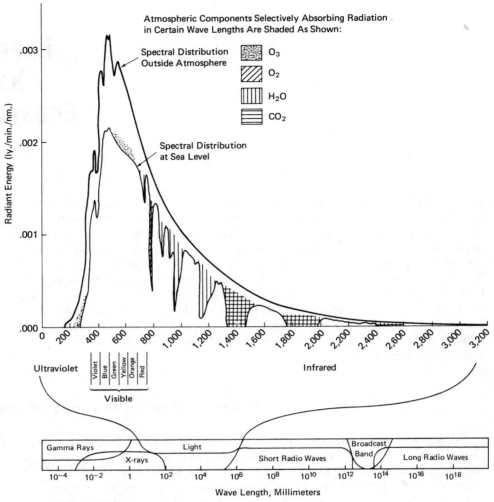

Figure 2.1

Spectrum of electromagnetic radiation from gamma rays to radio waves. Light spectrum is shown separately at sea level and at the outer edge of the atmosphere. When specific gases in the atmosphere have selectively absorbed certain wave lengths, the absorbing gas is indicated by shading. [Redrawn, in part, from US Air Force Cambridge Research Laboratories. *Handbook of Geophysics and Space Environments*. Copyright © 1965 by the United States Air Force.]

tions in the temperature zone. About 35% of the total is *albedo*, or reflectance back into space. High albedo would make the earth seem quite bright to an astronaut or resident of another planet, but light reflected into space is unavailable to any process in the biosphere. Roughly 24% reflects from the surfaces of clouds, 7% reflects from the gases in the atmosphere, and 4% reflects from the earth's surface itself. 14% is absorbed by particles in the atmosphere. When this happens, the energy is typically transformed into some other form of energy, generally chemical energy or heat. It is often reradiated as a less intense form of electromagnetic radiation, such as infrared. Direct insolation account for about 30% of the total, of which 4% is part of the earth's albedo. An additional 25% of the total bounces from particle to

Figure 2.2
Annual variation of insolation at the equator, 40°N., and 80°N. latitude. Values are in cal/cm² per min. [Redrawn, with permission, from G. R. Rumney, *Climatology and the World's Climates*. Copyright © 1968 by Macmillan Publishing Co., Inc., New York.]

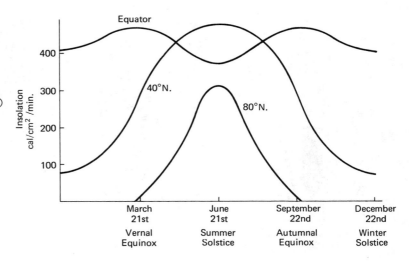

particle, so that it reaches the earth as diffuse visible light or as infrared reradiation from the atmosphere.

Spectral quality of insolation differs from the quality of light reaching the outer atmosphere. Almost all of the ultraviolet radiation is absorbed by the ozone layer in the stratosphere. This is vitally important to life, because ultraviolet is extremely toxic, and intense ultraviolet radiation would make the possibility of life speculative at best if the ozone layer were not present. Likewise, oxygen, ozone, carbon dioxide, and water vapor absorb much of the infrared radiation in certain bands. Much of this is reradiated as less energetic infrared; the rest is retained as heat. When energy is reradiated, it may also be absorbed by the atmosphere. Indeed, light may be absorbed and reradiated several times in the atmosphere. Each time, the reradiated light is less energetic than the light absorbed, and some energy is retained in the atmosphere as heat. This phenomenon is termed the *greenhouse effect*. It is important in heating the total atmosphere.

Figure 2.3
Disposition of solar radiation in the atmosphere. [Reprinted, with permission, from G. R. Rumney, *The Geosystem: Dynamic Integration of Land, Sea, and Air.* Copyright © 1970 by the Wm. C. Brown Company, Publishers, Dubuque, Iowa.]

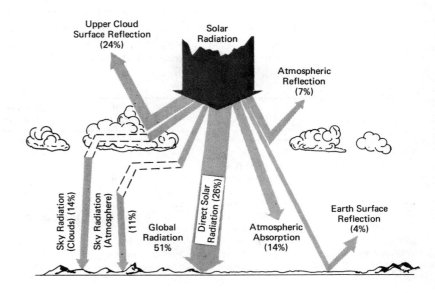

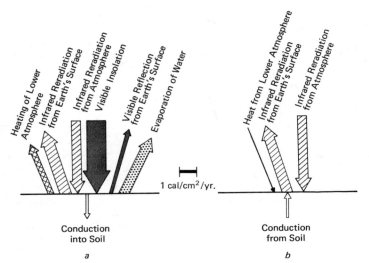

Figure 2.4
Energy balance at the earth's surface in Hamburg, Germany, in summer (a) between noon and 1:00 P.M.; (b) between midnight and 1:00 A.M. Width of arrow indicates magnitude of energy flow. [Redrawn, with permission, from R. Geiger, *The Climate Near the Ground,* 4th ed. Copyright © 1965 by the Harvard University Press, Cambridge.]

We can construct the energy budget at the earth's surface, showing the disposition of all energy received as insolation. Figure 2.4a shows the energy components of the budget at midday; Figure 2.4b shows them at midnight. Table 2.1 shows the energy budget for the earth's surface at Hamburg, Germany, averaged over an entire year. Of the total insolation, 89% is lost immediately as electromagnetic radiation of either short wave length (visible) or long wave length (infrared). It is the 11% that remains, or 5.6% of the solar constant, that is available to do the work of the ecosystem.

Disregarding the amount of energy incorporated into living organisms, about 14% of the remainder, or 4900 cal/cm²/yr, is utilized to heat the lower atmosphere. This is the main factor in insuring that the temperature of the biosphere is suitable for life. This heat also establishes eddy currents in the atmosphere that control weather and climatic patterns. The remainder of the energy, or about 30,000 cal/cm²/yr, goes to evaporate water. During the summer, the soil and the surface water may show a net surplus of heat, but almost all of this is lost during the winter. In addition, some parts of the earth generally show a slight heat surplus for the year, but this is balanced by other areas (mainly the polar icecaps) which show net heat deficits. Heat is transported from net surplus areas to net deficit areas by continental weather systems and global ocean circulation. Very little solar energy is stored permanently on the earth.

Table 2.1
Energy Budget at the Earth's Surface in Hamburg, Germany, 1953–54

Reprinted, with permission, from R. Geiger, *The Climate Near the Ground,* 4th ed. Copyright © 1965, by the Harvard University Press, Cambridge.

Category	Type of Light	Magnitude (in cal/cm²/yr)
Direct Insolation	Visible	34,153
Diffuse Insolation	Visible	43,444
Reradiation from Atmosphere	Infrared	240,533
TOTAL INSOLATION		318,130
Reradiation from Earth's Surface	Infrared	268,837
Reflection from Earth's Surface	Visible	14,367
TOTAL ENERGY LOSS FROM SURFACE		283,204
NET GAIN OF RADIANT ENERGY BY EARTH		34,926

Energy is neither created nor destroyed as light passes through the atmosphere; however, it may be changed from one form into another. These are the two principles of the *first law of thermodynamics*. Light energy can be transformed into the chemical energy of a green plant, energy of motion (kinetic energy), or heat. These forms of energy can be converted back into electromagnetic radiation. Or the light energy absorbed by a particle can be reradiated in part as electromagnetic radiation with a lower intensity and the remainder retained as heat.

The first law of thermodynamics is also known as the law of conservation of energy. It applies generally to energy conversion. But practical considerations of energy are seldom concerned with energy conversion as such. We are more commonly interested in doing specific things with energy. It is more important to know how much useful work is done by the energy input than to know that the sum of all the energy outputs (whether useful or not) is equal to the energy input. Efficiency of the conversion process is the keyword of practical energy application. This is the realm of the *second law of thermodynamics*, which states that the efficiency of energy conversion to useful work is never perfect; when energy changes from one form to another, some of the energy is rendered unavailable to do useful work in the system. It is useless heat, or *entropy*. It is not destroyed; it is still there. It just can't be used for anything. The second law of thermodynamics holds for all energy transformations, including those involving the biochemical energy of life.

To take an example, light energy may be absorbed by a molecule of oxygen in the atmosphere and transformed into kinetic energy, splitting the molecule into two atoms of oxygen. Such a reaction involving light is a *photochemical* reaction. When these oxygen atoms rejoin, they release their energy as electromagnetic radiation. Designating light energy as hv, entropy as S, and the energy incorporated in the two atoms of oxygen as W, we can follow the energy transfer through the splitting and rejoining of the oxygen atoms:

$$hv_1 \longrightarrow W + S_1 \qquad O_2 \longrightarrow 2O \quad \text{splitting of } O_2 \text{ molecule}$$
$$W \longrightarrow hv_2 + S_2 \qquad 2O \longrightarrow O_2 \quad \text{reformation of molecule}$$

There are two energy transformations. Some of the energy is lost as entropy at both stages $(S_1 + S_2)$. Thus the light energy reradiated by the oxygen atom (hv_2) is less than that originally absorbed by it (hv_1). This is the reason that reradiated light is in the infrared rather than the visible light band because infrared light contains less energy than visible light.

Energy in the Living Community

Thus far we have ignored life in our discussion of the earth's energy budget. In a way this is not unreasonable. Only a very small proportion of the total insolation is transformed into biochemical form. Several studies have been carried out to determine this percentage. One of the earliest indicated that of the total insolation on an Illinois cornfield, 1.6% was actually utilized by the corn.[1] A study of Lake Mendota in Madison, Wisconsin, reported that 0.39% of the insolation was utilized by aquatic organisms.[2] Vegetation in Cedar Lake Bog, Minnesota, utilizes about 0.10% of the energy reaching the surface.[3] In other words, the amount of energy needed to sustain life seems (at first glance, at any rate) extraordinarily small. The biological communities that are most effective in obtaining energy seldom incorporate more than 3% of the total insolation. We can use 1% as a rule of thumb for a terrestrial ecosystem.

This 1% is not as small a figure as it might seem. Remember that about 89% of the insolation is unavailable because it is lost as electromagnetic radiation. Furthermore, life is only one function of the ecosystem. If a plant community in Hamburg, with a total insolation of over 318,000 cal/cm²/yr could utilize 1% of this energy, it would convert 3,180 cal/cm²/yr into living tissue. This figure is only slightly less than the amount of energy available to heat the lower atmosphere. More interesting, perhaps, is a comparison between the energy utilization by a natural ecosystem and a cultivated crop. Intensively cultivated crops can convert as large a proportion of insolation into biomass as almost any ecosystem. But traditional agriculture tends to be much less productive. On a global average, cultivated cropland is less productive than most natural terrestrial ecosystems[4]. We have paid a high price to develop our current patterns of food production. We must put up with either low productivity or large "fuel subsidies": the use of fossil fuel energy to maintain or raise production.

Living organisms can use energy in several forms, but it is virtually all either radiant or fixed. Radiant energy is in the form of electromagnetic waves, such as light. Fixed energy is potential chemical energy embodied in various organic substances. It can be released when the substances are metabolized in living cells or burned. Organic substances are molecules containing the element carbon, and they are produced by living organisms. Organisms that can take energy from inorganic sources and fix it into energy-rich organic molecules are called *autotrophs* (from the Greek, "self-feeding"). The overwhelming majority are plants that obtain their energy directly from light; these are called *photosynthetic autotrophs*. Some bacteria obtain their energy from inorganic chemicals and are termed *chemosynthetic autotrophs*. Some very specialized organisms found in unusually warm areas of the very deep ocean appear able to utilize heat energy and sulfides that issue from deep fissures on the ocean bottom. Organisms that depend on energy-rich organic molecules (food) for their energy supplies are termed *heterotrophs* (from the Greek, "feeding on others"). There are two basically different types of heterotrophs. Those that obtain their energy from living organisms are called *consumers*, and those that obtain their energy either from dead organisms or from organic compounds dispersed in the environment are called *decomposers*.

Light energy is converted into chemical energy by *photosynthesis* in plants:

$$6CO_2 + 6H_2O + h\upsilon \longrightarrow 6C_6H_{12}O_6 + 6O_2$$

| atmospheric carbon dioxide | water from soil | light energy | sugar in plant cell | oxygen released to atmosphere |

The reaction is catalyzed by certain pigments in the cells, generally chlorophyll acting in concert with one or more other pigments. The product of this reaction is a carbohydrate (an organic compound composed of carbon, hydrogen, and oxygen) such as the sugar (glucose) shown in the preceding reaction. Several things can happen to it: It can be converted to a relatively inert energy-rich organic substance such as starch and then stored; it can be combined with other sugar molecules to form specialized carbohydrates such as cellulose, which are used by the plant for specific purposes; or it can be combined with other materials including nutrient substances such as nitrogen, phosphorus, and sulfur, as well as sugars and other simple carbohydrates, to build complex molecules such as proteins, nucleic acids, pigments, and hormones. All of these products are necessary for the normal growth and maintenance of the body tissues and functions of the plants. All require energy. Much of this energy is provided by using some of the sugar produced by photosynthesis as an energy source. It is combined with oxygen to give carbon dioxide, water, and usable chemical energy:

$$C_6H_{12}O_6 + 6O_2 \longrightarrow 6CO_2 + 6H_2O + \text{usable energy}$$

Oxidation of sugar (or any other organic chemical, for that matter) by organisms is called *respiration*. The energy released by respiration is lost permanently to the ecosystem.

Production and Productivity

Sugar is the primary product of photosynthesis, and all the sugar produced in the leaves of green plants is derived from carbon dioxide and water that have been combined by solar energy. Thus the energy incorporated into living tissue can be expressed in terms of either the light energy utilized or in terms of the sugar produced (Figure 2.5). It is not unusual to find studies of the energy relations of biological communities in which the amount of energy taken into the system is measured in kg. of sugar rather than a more normal energy unit such as calories.

Because all the energy fixed by the plant is converted into sugar, it is theoretically possible to determine plants' energy uptake by measuring the

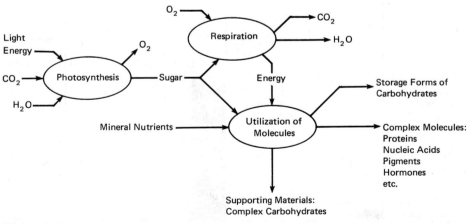

Figure 2.5
Diagram of the production of fixed carbon and its utilization as the building blocks of other materials and as a source of energy to power metabolism.

amount of sugar produced. This quantity is termed *gross primary production*. However, gross primary production is an abstraction that is difficult to measure. It *does* represent the sugar produced by photosynthesis, but a real plant needs energy to maintain its metabolism, and this energy is derived from photosynthetic sugar. Measurements of the buildup of sugar in the plant reflect gross primary production less respiration, or *net primary production*.

Measuring production is not an easy task; biomass is a relatively easy thing to measure. But biomass is a "snapshot" of the plant's energy relations, the aggregate of all the gross primary production and all of the respiration over the life of the plant. Production is a dynamic concept; it is a change in biomass over a period of observation. Measuring production means recognizing the significance of the time dimension in plant growth. Meaningful measurements of production depend on recognizing the factors that influence the rates of production and respiration and ensuring that comparisons of production rates in different ecosystems reflect similar conditions of growth.

To facilitate comparisons among ecosystems, most production estimates are taken for a specific interval of time, one year. The production of an ecosystem over a year is termed the *productivity*. As with production, productivity can be either net or gross.

Over a year's time, much can happen to the sugars produced by plants in a real ecosystem. Respiration accounts for the loss of much of the gross production. But a considerable proportion is consumed by herbivorous animals; some is embodied in tissues that die. Cropped and dead tissues no longer constitute part of the plant biomass and cannot be measured to calculate production. A complex and significant relationship between photosynthesis, respiration, cropping, and death is shown in Figure 2.6. Production estimates based on biomass measurements need to account for all of the losses of biomass. There are several techniques for doing this.[5] In addition, there are direct measurement techniques for production and respiration involving the rate of uptake of radioactive carbon-[14]C.

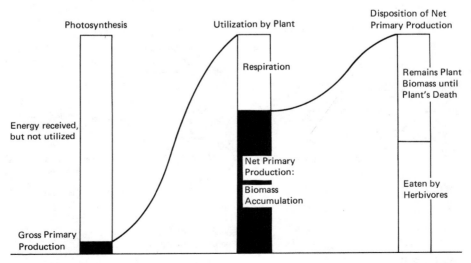

Figure 2.6
Disposition of energy received by a green plant.

Table 2.2
Estimated Annual Gross Primary Production of a Range of Ecosystems

Modified, with permission, from E. P. Odum, *Fundamentals of Ecology,* 3rd ed. Copyright © 1971 by the W. B. Saunders Company, Philadelphia.

Ecosystem	Area (10⁶ km.²)	Gross Primary Productivity (cal/cm²/yr)
Marine		
Open Ocean	326.0	100
Coastal Zones	34.0	200
Upwelling Zones	0.4	600
Estuaries and Reefs	2.0	2,000
Terrestrial		
Deserts and Tundras	40.0	20
Grasslands	42.0	250
Dry Forests	9.4	250
Boreal Conifer Forests	10.0	300
Cultivated Land (Unmechanized)	10.0	300
Moist Temperate Forests	4.9	800
Mechanized Agriculture	4.0	1,200
Wet Tropical, Subtropical Forests	14.7	2,000

Estimates of primary production in different environments yield some very interesting and consistent patterns. Table 2.2 lists estimates of the gross productivity of several different ecosystems. The geographic distribution of estimated annual net primary productivity for the world is shown in Figure 2.7. These figures are only an approximation of the distribution of productivity, but they are nevertheless useful. They underscore the tremendous differences between the productivities of different ecosystems, and they allow some creative generalizations. Warmer climates are characterized by greater production than cooler ones; wetter climates are more productive than dry ones. The most productive ecosystems are "open" ecosystems that have extensive communication with (and inputs from) other ecosystems; these include estuaries, swamps, rice paddies, and coral reefs. As a rule, semiclosed ecosystems, whose communication with adjacent ecosystems is minimal, are less productive. Deserts are very low in productivity, but, surprisingly, so also is most of the open ocean.

Primary productivity is the foundation for the metabolism of ecosystems. Because of this, the factors that determine the distribution of production have a key role in determining the structure of ecosystems. Indeed, some of the key questions that have impelled ecological research for the last 6 decades are: What factors control the fixation of light energy by green plants, how does the amount of fixed energy available in an ecosystem influence its structure, and how are the structure of the ecosystem and the patterns of energy passage related? We shall come back to these questions at several points throughout this book.

Food Chains

Biological communities include more than plants. They include *herbivores* (plant-eating animals), *carnivores* (meat-eating animals), and *detritivores* (organisms that can use the organic compounds in sediment and soil to power their metabolism). A useful, very simple model for following the flow of energy and materials through the community is the *grazing food chain* (Figure 2.8). Plant tissues can survive as such, they may be eaten, or

cal/cm² /yr.

over 320
240 – 320   "
160 – 240   "
80 – 160   "
40 – 80   "
0 – 40   "

Lands

0 – 20   "
20 – 40   "
40 – 80   "
over 80   "

Waters

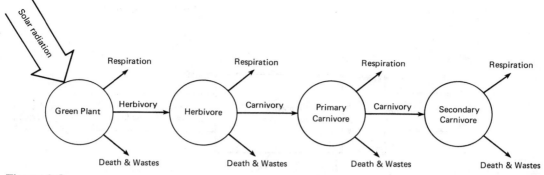

Figure 2.8
Passage of fixed energy along the grazing food chain.

they may die and decay (Figure 2.6). Herbivores gain their energy by consuming plants, and carnivores gain theirs from consuming other animals. We can describe energy flow in terms of these categories:

autotroph — herbivore — primary carnivore — secondary carnivore — etc.
primary producer   primary consumer   secondary consumer   tertiary consumer

The categories are functional levels linked to each other like links in a chain. This image of linkage is what gives the food chain its name. The levels themselves are often termed *trophic levels*, and they define an order for the passage of energy through the food chain. Like many very simple models, the notion of the food chain, with its associated trophic levels, provides a simple, meaningful abstraction to orient people to many features of energy flow through communities. It does not reflect the structure of real communities, but it epitomizes many of the processes that occur in them.

PRODUCTION BY CONSUMERS. Plant materials consumed by the herbivores are significant not only in that they represent biomass removed from plant populations but also because the energy content of that biomass is the theoretical maximum level of energy available to all animal populations—both the herbivores and the carnivores feeding in turn upon them. However, the actual amount of energy available to animal populations is not the gross amounts consumed by herbivores. It is rather the *assimilation* of energy by herbivore populations: the total plant consumption less the materials lost as feces. That is, it is the energy in the food that actually passes through the walls of the gut. This is sometimes termed *gross secondary production*.

Unlike gross primary production, gross secondary production can be measured directly by determining the amount of material ingested by the herbivore in question and subtracting the material defecated. As in plants, the food assimilated by animals can be stored as carbohydrate, protein, or fat. It can be transformed into relatively simple substances or rebuilt by the

Figure 2.7 (opposite)
Estimated annual net productivity of the various ecosystems of the world. Much of the figure is generalized. [Redrawn, with permission, from H. Lieth, "Versuch einer kartographischen Darstellung der Produktivität der Pflanzendecke auf der Erde," *Geographisches Taschenbuch*, 1964–1965, p. 72. Copyright © 1964 by the Franz Steiner-Verlag, G.m.b.H., Wiesbaden.]

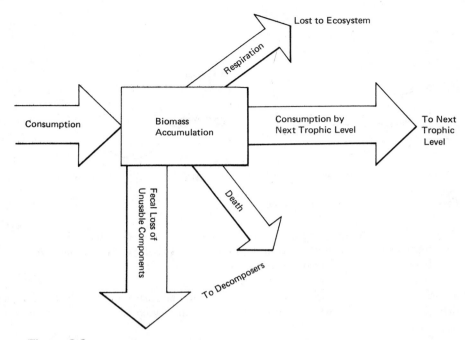

Figure 2.9
Disposition of energy consumed by herbivores.

animal into much more complex organic molecules. The energy to perform these transformations is supplied by respiration. The ultimate disposition of the energy assimilated by herbivores is by four routes: respiration, biomass accumulation, decay of dead organic matter by bacteria and other decomposer organisms, and consumption by carnivores (Figure 2.9).

Carnivores are animals that eat other animals. Those whose primary food source consists of herbivorous animals are termed *primary carnivores*, or secondary consumers; those whose main source of food is primary carnivores are termed *secondary carnivores*, or tertiary consumers; and so on. The energy consumed by primary carnivores is the energy content of the herbivores killed and eaten. As with herbivores, the energy assimilated by the carnivore (or *gross tertiary production*) is equal to the amount consumed less the losses to feces and other wastes. Its disposition into respiration, biomass accumulation, decay, and further consumption by other carnivores is entirely analogous to that of the herbivores.

The energy theoretically available for consumption by higher trophic levels declines rapidly as one goes up the food chain. The maximum available energy is the assimilation, but respiration takes much of this off the top, and some organisms die or lose tissues to decay prior to death (e.g., plant roots and leaves or the antlers of deer and their relatives). Other energy is lost by excretion. Biomass accumulation represents the amount of energy stored in a form available for consumption by higher trophic levels. But the disappearance of biomass would be equivalent to the disappearance of the population, so not all of the biomass pool is available to higher consumers. The decline in gross production as one goes up the trophic scale is shown by Table 2.3.

The decline in productivity as one rises in the food chain leads to a significant feature of community structure: The biomass at higher trophic levels is

Table 2.3
Gross Production for Several Trophic Levels in Several Different Ecosystems
(in cal/cm²/yr)

Ecosystem	Autotrophs	Herbivores	Primary Carnivores	Secondary Carnivores
1. Lake Mendota, Wisc.	480	41.6	2.3	0.3
2. Silver Spring, Fla.	20,810	3,368	383	21
3. Cedar Lake Bog, Minn.	120	16.8	3.1	—
4. Salt Marsh, S.C.	36,380	767	59	—

1. from Juday, 1940, as corrected by Lindeman, 1942; 2. from H. T. Odum, 1957;
3. from Lindeman, 1942; 4. from Teal, 1962.

usually significantly less than that at lower levels, and this is also commonly reflected in the numbers of organisms. The progressive decreases in numbers, biomass, and production is portrayed graphically in Figure 2.10. The decline in productivity is a general rule in communities, but there are exceptions to the rules of decreasing numbers and biomass. Small parasites, for example, may be much more common than their hosts, but production per parasite is so much less than that of the host that total parasite production is substantially less than total host production. A mature forest may contain many fewer trees than herbivores, but plant production and biomass are higher than those of herbivores because of the higher production and biomass of the individual tree. In many aquatic ecosystems, single-celled plants are less common than the tiny herbivores feeding on them, but they reproduce so much faster than the animals that their gross production is significantly higher.

Even the progressive decrease in production, though always true in a general sense, is not always as straightforward as indicated here. Many ecosystems are characterized by imports of substantial amounts of energy in the form of dead organic matter, or *detritus*. Consumer production may be higher than that of autotrophs in these ecosystems. Indeed, there are extreme cases, such as caves or deep lakes and seas below the penetration level of light, where there are no autotrophs at all. These ecosystems are subsidized by energy from external sources. Understanding their energy dynamics requires consideration not only of primary production within the community but also of the energy inputs from outside. Some of our most productive ecosystems are of this sort.

One indicator of the order in an ecosystem is the efficiency of energy transfer from one trophic level to the next. The most meaningful measure of this, and the one that has come into greatest use, is the *progressive effi-*

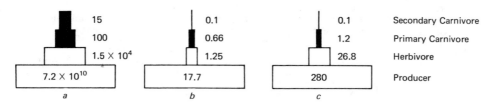

15	0.1	0.1	Secondary Carnivore
100	0.66	1.2	Primary Carnivore
1.5 × 10⁴	1.25	26.8	Herbivore
7.2 × 10¹⁰	17.7	280	Producer
a	b	c	

Figure 2.10
Community pyramids for a community in a shallow experimental pond. *(a)* pyramid of numbers, measured in individuals per m.²; *(b)* pyramid of biomass, measured in g. dry weight per m.²; *(c)* pyramid of productivity, measured in mg. dry weight per m.² per day. [Reprinted, with permission, from R. H. Whittaker. *Communities and Ecosystems.* Copyright ⓒ 1975 by Macmillan Publishing Co., Inc., New York.]

Cultivated Cornfield, Ill.	23.5%
Old Field, Southern Mich.	15.1
Broomsedge Field, S.C.	48.1
Root Spring, Mass.	7.75
Lake Mendota, Wisc.	22.3
Cedar Lake Bog, Minn.	21.0

Table 2.4
Autotroph Respiration in Several Ecosystems, Shown As a Percentage of Gross Primary Production

ciency or *gross ecological efficiency*. This is the ratio of gross production of any trophic level to the gross production of the trophic level preceding it. A survey of the ecological literature indicated that this ratio was on the order of 10% through the food chain and over a wide range of communities.[6]

RESPIRATION. Respiration is the process by which organic substances (generally simple sugars) are oxidized to carbon dioxide and water. It releases energy that can be used for biomass accumulation (growth), repair of old tissues, motion, and any other essential operation of the organism. Some of the energy released by respiration is incorporated into the complex organic substances built by the organism. The rest is lost to the ecosystem as heat. Respiratory energy conversion efficiency in living organisms is typically fairly low, and a great deal of energy is lost. Indeed, the amount of energy lost from any ecosystem through respiration at any trophic level typically constitutes a significant proportion of the gross production.

As might be anticipated, the respiratory losses from plants in different communities are highly variable. Estimates range from less than 10% to more than 75%, depending on the community and the environment in which it exists. Some examples of autotroph respiration as a percentage of gross primary production are shown in Table 2.4.

It is difficult to measure the respiratory losses of animals in the field because they move from one place to another. The measurements that have been made show the expected variation from ecosystem to ecosystem, but there do seem to be definite patterns (Table 2.5). Many communities show an increase in respiratory loss at higher trophic levels. This may be related to the energy expended by consumers in obtaining their food.

LIMITS TO THE LENGTH OF THE GRAZING FOOD CHAIN. How many trophic levels can a food chain contain? The ability of a species to exist at a given trophic level depends on its ability to assimilate sufficient energy to offset the losses to respiration and predation and to allow normal growth,

Table 2.5
Respiratory Losses at Several Trophic Levels in Several Ecosystems, Shown As Percentage of Food Assimilation (Total Ingestion Minus Defecation for Animals, Gross Production for Plants)

Ecosystem	Respiratory Losses			
	Autotrophs	Herbivores	Primary Carnivores	Secondary Carnivores
1. Lake Mendota, Wisc.	22.3	36.1	47.8	66.7
2. Root Spring, Mass.	7.75	75.4	36.8	—
3. Silver Spring, Fla.	57.5	56.1	82.5	61.9
4. Cedar Lake Bog, Minn.	25.0	38.1	58.1	—
5. Salt Marsh, S.C.	77.5	77.7	81.3	—

1. from Juday, 1940, as corrected by Lindeman, 1942; 2. from Teal, 1957; 3. from H. T. Odum, 1957; 4. from Lindeman, 1942; 5. from Teal, 1962.

tissue maintenance, metabolism, and reproduction. Because of the average progressive efficiency of ecosystems, the total energy available to the organisms of any trophic level decreases exponentially: Each trophic level must make do with about 10% of the energy available to the next lower level. If energy requirements for the organisms at each trophic level also decreased exponentially at a similar rate, the food chain could theoretically extend indefinitely. But they do not. If anything, the increase in respiratory loss as a function of production indicates that energy requirements increase going up the food chain.

The number of trophic levels that can be maintained in any ecosystem is finite and small. The limit is reached when animals can no longer assimilate sufficient energy to balance their energy expenditures. This may be at the primary carnivore level in small ecosystems. Only rarely are ecosystems sufficiently productive and stable to have more than five trophic levels; most familiar communities have about four.

DEATH AND WASTES. A substantial amount of organic material can be lost from one trophic level without being assimilated by the next higher level. For example, animal feces represent material ingested but not assimilated. Defecation as a proportion of overall consumption is exceedingly variable, ranging from very low to as high as 90%.[7] In the same way, exudates of various sorts are excreted by the leaves and roots of most plants. Many animals and plants routinely shed tissues, such as old roots and leaves from plants or skins from molting animals such as snakes and arthropods. Finally, dead animals and plants that are not totally eaten by consumers are lost to the grazing food chain.

Most of the organic materials that fall into this category are rich in energy and can serve as the basis of the metabolism of other organisms. Materials lost to the grazing food chain are not lost to the ecosystem. They are processed by a different set of organisms that most of us are much less familiar with: these are the *detritus food chain* (Figure 2.11).

THE DETRITUS FOOD CHAIN. The detritus food chain represents an exceedingly important component in the energy flow of an ecosystem. Indeed, considerably more energy flows through the detritus food chain in some ecosystems than through the grazing food chain. The detritus food chain differs from the grazing food chain in several ways. First, the organisms making it up tend to be smaller. Second, the functional roles of different populations do not fall as neatly into trophic levels. Finally, detritivores inhabit an environment that surrounds them with dispersed food particles. As a result, energy storage may be in the detritus itself, largely external to the organisms.

Detritivores typically constitute a melange of many different types of organisms with somewhat different but overlapping roles. It is most meaningful to visualize energy flow in the detritus food chain as a continuous passage rather than as a stepwise flow between discrete entities. The organisms of the detritus food chain include members of many phyla of animals and plants, such as algae, bacteria, slime molds, fungi, protozoa, insects, mites, crustaceans, centipedes, mollusks, worms, sea cucumbers, and even some vertebrates. Some species are highly specific in their food requirements, whereas others can eat almost anything. Many protozoa, for instance, need

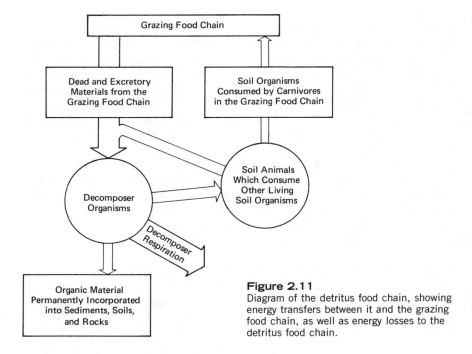

Figure 2.11
Diagram of the detritus food chain, showing energy transfers between it and the grazing food chain, as well as energy losses to the detritus food chain.

certain specific organic acids, vitamins, and other nutrients in order to thrive. On the other hand, the guts of some Collembola (a group of tiny soil insects) have been reported to contain decaying plant material, fungal fragments, spores, fly pupae, other Collembola, parts of decaying earthworms, and cuticle from their own fecal casting.[8] Some soil organisms will ingest only certain types of matter; others will voraciously gobble their way through the soil, consuming anything that happens to be present, including, in some laboratory experiments, such tidbits as charcoal and plaster of Paris.

Many detritus organisms ingest pieces of partially decomposed organic matter, partially digest them, and excrete the remainder as slightly simpler organic molecules, after extracting some of the energy in the detritus to run their metabolism. The wastes from one organism can immediately be utilized by another, which repeats the process (Figure 2.12). Gradually, the complex organic molecules present in the original wastes or dead tissue are broken down to much simpler compounds, sometimes all the way to carbon dioxide and water. In most instances, however, the organic material is broken down until all the easily biodegradable material has been degraded and all that is left are some refractory organic substances that are incorporated

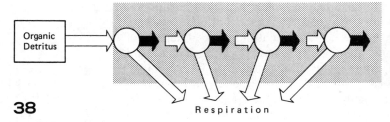
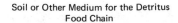

Figure 2.12
Passage of energy through the detritus food chain. Each decomposer organism can utilize the waste products of previous decomposer organisms as food.

into the soil or sediment. These substances can be exceedingly stable. Radio-carbon dating of humic acids from several soils shows mean longevities as high as 400 to 2860 years.[9]

Other detritus organisms obtain the greater part of their energy needs not from the detritus, but rather from other detritus organisms. Many animals prefer bacteria as a food source, even if they can assimilate dead tissue or partially decomposed materials. It is no wonder that it is so difficult to assign functional roles within the detritus food chain.

The areas where the detritus food chain operates in the soil and in aquatic ecosystems are full of physical activity, even if the small size of the organisms involved disguises this fact. Much of the dynamism of this food chain depends on physical movement. For example, oxygen is introduced into terrestrial soils, so that detritivores can obtain energy by oxidizing organic substances, because earthworms plow through the soil and aerate it. In addition, earthworm feces are richer in mineral nutrients and organic matter than the rest of the soil, and they also have a very high microorganism content which acts to break down detritus faster than in the absence of earthworms. In addition, the larger detritus organisms, insects, and mites, physically move detritus particles such as leaf litter. This mixing is essential for the efficient operation of the detritus food chain.

The detritus food chain can extract a prodigious amount of energy. Decomposer organisms are often exceedingly active, processing large amounts of matter and releasing a great deal of energy mostly as heat. (This can be seen dramatically by comparing the temperature of a compost heap with the air temperature around it. The compost heap, a cultured detritus-based ecosystem, will be warmer.) Respiration by the detritus food chain typically releases a large proportion of the energy that comes into it as detritus. In H. T. Odum's 1957 study of Silver Springs, Florida, 5060 kcal/m² of detritus were provided annually, of which 4600 kcal/m², or 90.8%, was released from the ecosystem through respiration. Detritus breakdown in mature forest soils tends to bring about an equilibrium between litterfall and breakdown. The two are equal over an annual cycle, and the amount of litter in the soil stays constant.

The efficiency of the detritus food chain depends on many factors, of which the most important are the oxygen content of the medium and certain gross environmental variables such as temperature and rainfall. If oxygen is consistently available, energy is obtained through aerobic respiration, as it is by the organisms of the grazing food chain. However, oxygen may not be readily available in many environments where detritus is being consumed; for instance, within the sediments at the bottom of a lake, in a waterlogged soil such as those of marshes and swamps, or in the interior of a decaying carcass. When free oxygen is unavailable, aerobic respiration cannot take place, and the breakdown of organic compounds and energy extraction must then proceed by *anaerobic* means (e.g., fermentation). Large organic molecules are split into smaller ones without the addition of oxygen. Examples are the fermentation of sugar to form alcohol and carbon dioxide or to form lactic or pyruvic acid (Figure 2.13).[10]

The energy released by anaerobic breakdown of organic materials is sufficient to drive the metabolism of some organisms (generally bacteria, yeasts, and other simple microorganisms). But anaerobic reactions release much less energy from organic materials than does aerobic respiration, as shown in

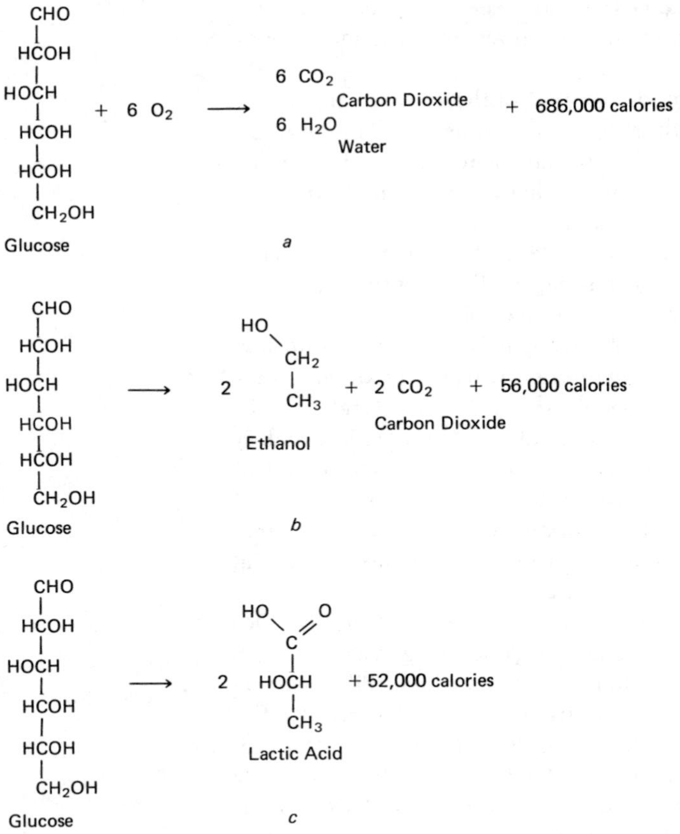

CHO
|
HCOH
|
HOCH   6 CO_2
| + 6 O_2 →   Carbon Dioxide   + 686,000 calories
HCOH   6 H_2O
| Water
HCOH
|
CH_2OH

Glucose   a

Figure 2.13
Comparison of the amount of energy released through fermentation and respiration of one mole (180.15 g.) of glucose: (a) respiration of glucose to its basic constituents. (b) fermentation of glucose to ethanol, as in a yeast culture; (c) fermentation of glucose to lactic acid, as in a muscle. Note that a given amount of glucose will produce over twelve times as much energy from respiration as it will from fermentation.

CHO
|
HCOH   HO
| \
HOCH CH_2
| →   2 | + 2 CO_2 + 56,000 calories
HCOH CH_3 Carbon Dioxide
|
HCOH   Ethanol
|
CH_2OH

Glucose   b

CHO
|
HCOH   HO   O
| \ //
HOCH   C
| →   2 | + 52,000 calories
HCOH   HOCH
| |
HCOH CH_3
|
CH_2OH Lactic Acid

Glucose   c

Figure 2.13. End products of anaerobic breakdown include relatively inert molecules such as carbon dioxide, water, and humus. They also include small-molecule alcohols, organic acids, ptomaines, amines, and other products, as well as gaseous substances such as methane (CH_4). Some of these may be quite toxic in high concentration, and they may accumulate in areas in which they are produced instead of being degraded. Many also smell bad and confer a distinctive bouquet to ecosystems in which they are produced, such as a marsh or a polluted river. From an energy standpoint, volatile compounds that enter the atmosphere represent a net energy loss to the ecosystem. Soluble compounds which enter waterways are generally oxidized by aquatic organisms, but the concentration of these soluble wastes is so high in some lakes that virtually nothing will grow: bogs represent this type of environment.

The differences between aerobic and anaerobic decomposition as the energy extraction mechanism for the detritus food chain are significant in natural ecosystems, but they are even more so for ecosystems that have been altered by people. First, because respiration is so much more efficient at releasing the energy contained in organic molecules, the activity of the detritus food chain is much higher in an aerobic environment, and breakdown of materials is much more complete. Indeed, in some environments, such as warm, wet forests, essentially all the detritus available is completely oxidized to carbon dioxide and water. In ecosystems less optimal for the growth of bacteria, the refractory organic molecules remain as humus after most of the material has been degraded. At the same time, the nutrient elements, such as nitrogen, phosphorus, and sulfur, are released into the environment where they can be reutilized by living organisms. Detritus breakdown is substantially slower and less complete in anaerobic environments, however, resulting in the accumulation of undegraded detritus in the form of peat, organic soils, and highly organic sediments.

PERMANENT STORAGE OF FIXED CARBON IN SEDIMENTS. Some refractory organic materials remain in sediments and soils. They may constitute a relatively small proportion of the soil, as in the low organic soils of the wet tropics; they may be very abundant, as in peat bogs; or they may lie anywhere in between. This buildup of organic material in sediments represents the storage of solar energy more or less permanently within the sediments. Because the average annual increment of permanently stored organic fixed carbon over the surface of the earth is about 0.03 g/m^2,[11] this energy gain is significant over a period of time. Indeed, it is the source of all our fossil fuels, including coal, oil, and gas. Even more significant, humic materials in sedimentary shales account for roughly 600 times the amount of fixed carbon stored in the fossil fuels.[12]

The Earth's present oxygen-rich atmosphere is the result of this permanent storage of carbon, coupled with short-term storage of organic material in biomass. Photosynthesis in the Earth's primeval oceans combined reduced carbon in the then oxygenless atmosphere with water to make living tissue and release oxygen into the atmosphere. Photosynthesis produces six molecules of oxygen for every molecule of sugar. However, oxidation consumes all this oxygen when the sugar is respired. Thus, the presence of oxygen in the atmosphere depends on a mechanism for permanently storing fixed carbon in the earth's crust in a reduced form.

LINKS BETWEEN THE GRAZING AND DETRITUS FOOD CHAINS. The line between the two food chains is not always as sharp as in textbook discussions. Certain animals perform a "bridging" function and tie the energy flow patterns of an ecosystem much tighter than they might otherwise be. For example, many aquatic animals, including many clams, fish, and smaller invertebrates are "filter-feeders." They move water over specialized organs which remove suspended food particles. Most filter-feeders can use a broad range of food types, including detritus and small animals and plants. Even if these animals are functional detritivores (i.e., a substantial proportion of their food supply is detritus), they are preyed upon as though they were herbivores. Other animals, such as earthworms, are significant members of the detritus food chain, but they are preyed upon by robins. Many insects, including some beetles and flies, spend their larval period as detritivores and their adulthood as herbivores or carnivores. Links such as these, by which energy passes from the detritus food chain back into the grazing food chain, may be locally important.

Such linkages are especially important for ecosystems which depend on import of energy from outside. In many environments, photosynthesis produces much less fixed carbon than is consumed by animals. These include environments in which no photosynthesis occurs at all, such as caves and deep water below the penetration limit of light, as well as places where plants have difficulty rooting themselves, such as streams. Many of our most

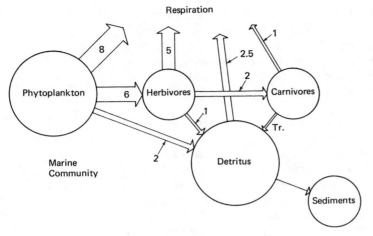

Respiration

Figure 2.14
Energy flow model for marine and forest communities. Energy expressed in cal/cm²/10-day period. Some numbers may not add up, because of rounding. [After Odum, 1963.]

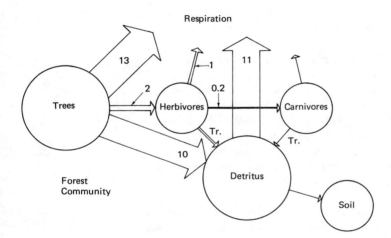

Respiration

productive environments, in fact, depend largely on energy imports from other ecosystems. The imports are in the form of detritus, which is consumed by an animal (generally a filter-feeder) that is then consumed by a member of the grazing food chain.

RELATIVE ROLE OF THE TWO FOOD CHAINS. It is a common perception that the grazing food chain, with its larger organisms, is somehow more "significant" than the detritus food chain. Let us, therefore, examine the

Table 2.6
Energy Disposition from Grazing and Detritus Food Chains in Marine and Forest Ecosystems (in cal/cm²/day)

Data from Odum, 1963.

Ecosystem	Net Primary Production	Gross Secondary Production	Gross Detritus Production
Marine	0.8	0.6	0.2
Forest	1.2	0.2	1.0

	Gross Heterotroph Respiration	
	Grazing Food Chain	**Detritus Food Chain**
Marine	0.5 (63%)	0.3 (37%)
Forest	0.1 (8%)	1.1 (92%)

energy dynamics of each in some representative ecosystems. The comparisons are intriguing. Figure 2.14 shows respiration and energy transfer in a marine community and a forest community. The figures represent the magnitude of the estimated energy flow within the communities and are summarized in Table 2.6. The grazing food chain accounts for relatively more of the energy flow in the marine ecosystem than in the forest, whereas the detritus food chain accounts for more in the forest. In both cases, the detritus food chain accounts for a large proportion of the total respiration.

The Food Web

The food chain provides a useful model for orienting somebody to the basic principles of energy flow in ecosystems. But just as the boundary line between the grazing and detritus food chains becomes fuzzy when one tries to find it in nature, the food chain model itself often leaves much to be desired. The model is based on trophic levels, which refer in a very abstract way to functions of organisms with regard to energy flow. But real species have a role in the community that relates to many different factors and that has been fixed in the population's genetic base through natural selection. Energy flow is only one of these factors. Some species may fit readily into specific trophic levels, but many do not.

In order to describe the actual flow patterns for energy in a community, it is most useful to invoke a model that is based on the food chain but considers the interactions among the populations making up the community in question. This is the *food web*. Populations are arrayed as completely as possible into trophic levels and are connected with arrows indicating the flow of energy from one population to another (i.e., through consumption of the one by the other). An example, from a simple New England meadow pond, is shown in Figure 2.15.

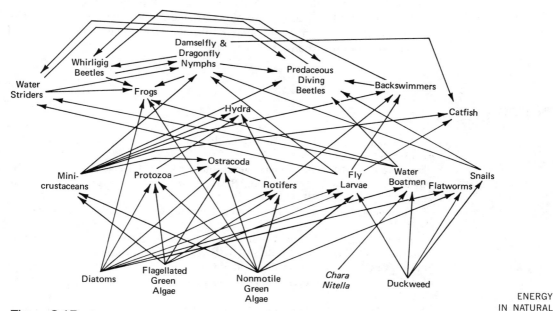

Figure 2.15
Diagrammatic representation, highly simplified, of the food web of a small meadow pond. Arrows point in the direction of energy flow.

Most of the organisms in this example fit into trophic levels, but some do not. The organisms most consistent in their trophic levels are green plants; few of them consume other organisms. (Some autotrophs, however, do gain significant quantities of their nutrients from other plants, or even animals. These include certain plants like the mistletoe, whose habitat is the upper branches of larger trees, as well as the insectivorous plants such as the Venus' flytrap and the pitcher plant. With regard to energy, all of these plants are autotrophs; with regard to nutrients, they are partially "herbivores" or even "carnivores"). The herbivores are less consistent. Most eat only plants (except when they ingest an animal by mistake, of course). Others, like the ostracod, regularly eat animals if they can catch them, although their usual diet consists of plants and detritus. Carnivores are still less consistent. Some species are predominantly primary carnivores; that is, they eat mainly herbivores. Others are mainly secondary carnivores, with a diet composed largely of primary carnivores, and so forth. But few carnivores will avoid eating any animal they can catch if they are hungry.

Most ecosystems have prominent species that do not fit neatly into trophic levels. Some large animals, such as raccoons, bears, and people, are omnivores that feed at all trophic levels in the community. Others, such as frogs and many insects, occupy different trophic levels at different stages in their life cycles. Does this mean that the trophic levels are meaningless? Clearly not. Calling something a primary or secondary carnivore may be an excellent way of describing of its primary adaptation. It provides a basis for comparing it with other organisms with similar basic adaptations. But it is likely to be only a general description of what a population actually eats.

The food web shown in Figure 2.15 demonstrates that the patterns of energy flow are tremendously flexible even in a tiny community. Animals have some choice of food supply, some of them a great deal. As a result, the adaptations of the different species provide an interlinked system that is much more complex than the simple food chain model would suggest. The signals linking the elements of the community may be far more varied and complex than would seem at first glance.

Notes

[1] Transeau, E. N., 1926. The accumulation of energy by plants. *Ohio Jour. Science*, **26**, 1–10.

[2] Juday, C., 1940. The annual energy budget of an inland lake. *Ecology*, **21**, 438–450, as corrected by Lindeman, R. L., 1942. The trophic-dynamic aspect of ecology. *Ecology*, **23**, 399–418.

[3] Lindeman, R. L., 1942. The trophic-dynamic aspect of ecology. *Ecology*, **23**, 399–418.

[4] Miller, D. H., 1965. The heat and water budget of the earth's surface. *Adv. in Geophysics*, **11**, 175–302; Whittaker, R. H., 1975. *Communities and Ecosystems*, 2nd ed. New York: Macmillan Publishing Company.

[5] Odum, E. P., 1968. Energy flow in ecosystems: A historical review. *Am. Zool.*, **8**, 11–18.

[6] Kozlovsky, D. G., 1968. A critical evaluation of the trophic level concept, I: Ecological Efficiencies. *Ecology*, **49**, 48–59.

[7] Phillipson, J., 1966. *Ecological Energetics*. New York: St. Martin's Press.

[8] Hale, W. G., 1967. Collembola. Chap. 12 in Burges, A., and Raw, F., *Soil Biology*. London: Academic Press Ltd., 397–411.

[9] Burges, A., 1967. The decomposition of organic material in soil. Chap. 16 in Burges, A., and Raw, F. *Soil Biology*. London: Academic Press Ltd., 479–492.

[10] Wald, G., 1963. Origin of life. In National Academy of Science, *The Scientific Endeavour*. New York: Rockefeller University Press.: 113–134; Broecker, W. S., 1970. Man's oxygen reserves. *Science*, **168**, 1537–1538.

[11] Borchert, H., 1951. Zur Geochemie des Kohlenstoffe. *Geochim. Cosmochim. Acta*, **2**, 62–75.

Materials Movement and Cycles in Ecosystems

3

ENERGY, IN THE form of food or light, is only one of the many resource needs for living organisms. Organisms need water and certain critical elements to manufacture particular molecules that are essential for their functioning. They need an environment maintained within a relatively restricted range of temperature and salinity, and an appropriate habitat for growing, reproducing, and so on. There are significant differences between the dynamics of energy and materials in ecosystems, just as there are certain things in common.

Nutrient Resources and Limitations

Energy flows through the food web in one direction: It is fixed in chemical form by photosynthesis, and it is either degraded to heat as organisms utilize it to power their metabolism or stored permanently in sediments. But the materials of the biosphere consist of chemical elements such as carbon, oxygen, hydrogen, nitrogen, phosphorus, and sulfur. Whatever role they play in the ecosystem is due to their chemical nature. Unlike energy, chemical elements are not continuously supplied from an extraterrestrial source. They are not degraded into unusable form by metabolism, although they may change their chemical form. An analogue of the first law of thermodynamics applies to chemical elements: the law of conservation of mass. This law states that matter, like energy, is neither created nor destroyed. An atom of, say, carbon is an atom of carbon regardless of whether it is found in a piece of coal, a pesticide molecule, a fish in the deep ocean, or a human tongue. Indeed, several million carbon atoms have probably been part of all of these materials within the last few billion years. Materials are not lost; they cycle through the ecosystem, constantly being reused.

The critical factor for life is whether adequate supplies of the material are available to organisms. What does "adequate" mean, and what mechanisms do ecosystems have to make nutrients available to living organisms? The question of adequacy is best described in terms of gradients in environmental factors; the basic mechanism for making chemical materials available to living organisms is the biogeochemical cycle.

Gradients in Environmental Factors

There are two types of abiotic factors in any ecosystem: physical (e.g., temperature, quantity and spectral quality of light, depth of water, rainfall, climate) and chemical (e.g., pH, nutrient availability). The physical variables tend to be "given" for any particular ecosystem; that is they are set by the fundamental geographic setting of the ecosystem and are not modified extensively by the actions of living organisms. The chemical factors, on the other hand, are strongly influenced by the community, and the cyclic patterns of nutrients and other materials from the community to the abiotic environment and back reflect a close interplay between the living and nonliving elements of the ecosystem.

Most abiotic factors have a feasible range: There is a minimum level and a maximum level. For a physical factor such as temperature, the minimum temperature is the winter low, and the maximum is the summer high. A chemical factor like phosphorus concentration in a lake also has minimum and maximum concentrations that can exist in the lake. In either case, there is a gradient between the two, and the actual level of the factor (temperature, phosphorus concentration) can be anywhere within the feasible range.

No population is adapted to survive under all conditions present in all ecosystems. Just as ecosystems have maxima and minima that are feasible in their physical settings, populations have maxima and minima for their survival or for the performance of key activities, such as reproduction. Furthermore, they are typically better adapted to certain parts of the gradient than to others. The portion of the range of variation within which a species can survive and function is defined as the *tolerance range* (Figure 3.1). That level within the tolerance range at which the population can function most effectively is termed the *optimum*.

Because organisms have many different resource requirements, each population has a characteristic tolerance range and optimum for many different factors. Survival and reproduction of the population require that the species be matched to its environment. The feasible gradient for all factors must be within its tolerance range. Simply because a species can function over a given range of a specific variable does not mean that it will be found wherever that variable is within its tolerance range. If one factor (e.g., temperature) is optimal in an area, but another (e.g., available moisture) is outside the tolerance range, the species will not be found.

Optima and tolerance ranges can be determined experimentally by establishing an artificial environment that holds all variables constant except for the one being tested and charting the reaction of the species. The reaction being measured (such as rate of respiration, growth, motion, or reproduction) is generally related to the survival of the organism in some way. However, the responses of a population to the complex of factors found in natural ecosystems may be much more intricate than its response to a single factor

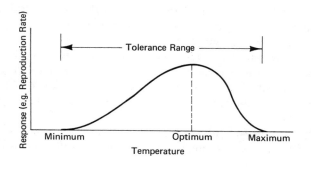

Figure 3.1
Temperature response curve for a typical animal, showing maximum and minimum temperature tolerance, tolerance range, and optimum.

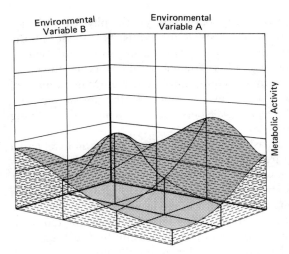

Environmental Variable B

Environmental Variable A

Metabolic Activity

Figure 3.2
Response curve of a given population to two environmental variables. Note that the optimum level for either of the two variables depends on the value of the other.

gradient. For example, rainfall is a critical factor for terrestrial plants, but the amount of moisture actually available to the plants from a given amount of rainfall depends not only on the amount of rain, but also on the ability of the soil to hold water, the temperature, and the other organisms in the community. The nutrition requirements of a species may vary with the temperature or the water supply. A population's requirements may change with time of year or stages in the life cycle. These factors can complicate the relatively simple notions of optimum level and tolerance range, as shown diagrammatically in Figure 3.2.

Sometimes, a species does not occur where one would expect it to. It is capable of colonizing the area in question, and the area is optimal with regard to certain key characteristics. But the species is found in other areas that seem, on the surface, to be less appropriate. Why? Obviously some factors are more important than others in determining the distribution of populations. We must distinguish, however, between the significance of a factor in controlling basic biological processes and the specific signals that control real populations.

An excellent example is the narrow-leaved cattail, *Typha angustifolia*, in eastern Nebraska.[1] The local race is restricted to salt flats with highly saline soil water. However, it can be shown to thrive under conditions of low salinity under experimental conditions. Nobody would deny the significance of the salinity of the soil solution to the survival of these cattails. In fact, *T. angustifolia* is not found in the freshwater moist areas because the broad-leaved cattail, *Typha latifolia*, is more successful than *T. angustifolia* at exploiting the less saline ecosystem, and the latter is simply forced out. Other instances of species being restricted to habitats that are suboptimal with respect to an important abiotic factor are not uncommon, especially for organisms normally found in harsh environments.[2]

In the same way, the marine mussel, *Mytilus californicanus* grows largest in the nearshore subtidal zone of the ocean, but it is seldom found there. It is much more common in the intertidal zone (between the limits of low tide and high tide), where it is not eliminated by predatory pressure from the starfish, *Pisaster ochraeceous*.[3]

Considering the complexity of the tolerance-response curves shown by any population for more than one environmental factor and the differences between the tolerance curve and the actual distribution of a species with respect to an environmental factor, one might wonder why we even worry about tolerance and optimum. The reason is simple: these concepts are real and describe important properties of the relationship between populations and environments. Organisms respond to every abiotic factor in the ecosystem. Some of these (especially those related to nutrient availability) are not as clear-cut as the salinity and temperature examples that one finds in textbooks. But the principle is the same. If an organism can theoretically exist in a particular environment and does not, it may be limited by factors that are not as obvious as those thought to be most important, or there may be barriers that prevent immigration to the site. Often, the factor we believe to be most important in controlling the distribution of a population turns out not to be so, and understanding why ecosystems are as they are requires detailed study of organisms' needs and the availability of several different materials.

Different species have different requirements. Some can survive under many levels of a particular variable (e.g., phosphorus concentration in a lake), whereas others can survive under only a very narrow range. Requirements may be broad for some factors and restricted for others. They may differ with stages in the life cycle or among different geographic races. For example, the requirements of a tadpole differ from those of a frog, as do those of a seed, a seedling, and an adult pine tree. Less obvious, but just as important, the reproductive requirements in most animals are more restrictive than requirements for mere survival. Many hawks and eagles, for example, could survive in areas in which DDT had been sprayed. But the DDT affected their calcium metabolism so that they could not lay viable eggs.[4] Other widespread species have several *ecotypes*, or geographic races, whose genetically based tolerance ranges and optima correspond to the environments within which they normally live. These are especially common in plants,[5] but they also exist in animals.

An example of genetically based differences in tolerance is shown in Figure 3.3. The temperature leading to death in two populations of bluegill (*Lepomis macrochirus*) was measured experimentally. The population from "warm water" (i.e. a pond heated by the effluent from nuclear reactors)

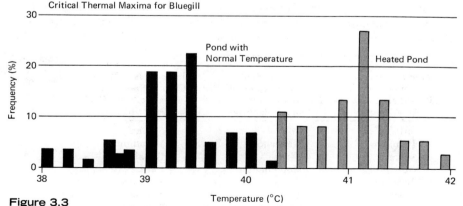

Figure 3.3
Critical thermal maxima for Bluegill (Lepomis macrochirus) taken from a pond with elevated temperatures and from a pond with normal temperature. After J. W. Gibbons and R. R. Sharitz, 1974. Thermal alteration of aquatic ecosystems. *Am. Sci.*, **62**:660–670.

could withstand significantly higher temperature than that from "normal water" (i.e., without the heated effluent).[6]

Limiting Factors

Justus Liebig noticed in 1840 that the population density of vascular plants could be limited if there was an insufficient amount of any important nutrient, and that the size of the crop could be increased if that nutrient were added to the soil. A factor whose variation has this kind of significance in determining population density is called a *limiting factor*. All environmental factors are potentially limiting, but the concept refers most often to nutrient availability and also to space.

Some 16 different chemical elements have been identified as essential for the survival of all species. These are the *essential elements*. Several other elements are needed in small quantities by some species (Table 3.1). Some, the *macronutrients*, are used in relatively large quantities as fundamental building blocks for organic tissues. Others, the *micronutrients*, are used in much smaller quantities. As an example, a corn crop will remove over 45 kg of nitrogen per acre from the soil but only 10 to 15 g of boron.

All species have tolerance ranges and optima for the nutrients they require, but few populations are lucky enough to inhabit areas where all nutrients are at just their optimal levels. Availability of materials in real ecosystems tends to be either consistently low or consistently high. For example, phosphorus is almost always present in suboptimal quantity, to such an extent that its dearth limits productivity. Selenium, on the other hand, is required by livestock and poultry,[7] but accumulates in certain plants to such a high concentration that it is highly toxic to these animals; selenium is more often superoptimal to the point of being poisonous than it is suboptimal.

Limiting factor analysis has been most useful in understanding how to control nutrient-driven pollution of lakes. Phosphorus is the most common limiting nutrient in fresh-water lakes. Its impact is shown in one of the experiments done in the Experimental Lakes Area of northwestern Ontario by the Fisheries Research Board of Canada. A small lake with two similar basins separated by a narrow neck was divided by means of a nylon-reinforced vinyl sea curtain. Nitrate and sugar (sucrose) were added to both basins in 20 equal weekly amounts. One basin also received phosphate along with the other inputs. The algal flora in the basin not receiving the phosphate was similar to that prior to fertilization, while the other basin underwent a tremendous bloom of the blue-green alga, *Anabaena spiroides* (Fig-

Macronutrients: Elements Used in Relatively Large Quantities	Micronutrients: Elements Generally Needed in Relatively Small Quantities	Micronutrients: Elements Needed by Certain Species in Relatively Small Quantities
Carbon	Iron	Sodium
Hydrogen	Manganese	Vanadium
Oxygen	Boron	Cobalt
Nitrogen	Molybdenum	Iodine
Phosphorus	Copper	Selenium
Potassium	Zinc	Silicon
Calcium	Chlorine	Fluorine
Magnesium		Barium
Sulfur		

Table 3.1
Essential Elements

Figure 3.4
Lake 226 in the Experimental Lakes Area of northwestern Ontario. The far basin has been fertilized with phosphorus, nitrogen, and carbon, while the near basin was fertilized only with nitrogen and carbon. The far basin was covered by an algal bloom within two months of the beginning of fertilization, while no increases in algal numbers or species were seen in the near basin. [Photograph courtesy D. W. Schindler; Freshwater Institute, Environment Canada. From Schindler, D. W. 1974. Eutrophication and recovery in experimental lake: implication for lake management. *Science* 184, 897–899. Copyright © 1974 by the American Association for the Advancement of Science, Washington.]

ure 3.4).[8] In this particular case, at least, phosphate availability limited plant growth, and the algal population was able to respond when the phosphate concentration increased.

In New Zealand, sheep developed a serious cobalt-deficiency disease termed bush sickness. It could be completely controlled by adding cobalt fertilizer to the range. These examples corroborate the observation that "when a process is conditioned as to its rapidity by a number of separate factors, the rate of the process is limited by the pace of the 'slowest' factor."[9] There can be no question that single factors often have a predominant effect on populations. However, the straightforwardness of the concept of limiting factors conceals a trap, because it implies a simplicity to the responses by populations to nutrient availability that does not always apply. It is significant that the limiting factor concept originated with, and is most clearly demonstrable with, very simple ecosystems such as agricultural fields or experiments involving only a few species. As will be shown in Chapter 6, a simple community tends to be more subject to perturbation by fluctuating abiotic variables than a complex community. Regulation of populations by single factors may depend as much on the complexity of the community as on any abiotic limiting factor. Thus it may be misleading to look for "the" limiting factor that explains the population size of any species in a natural ecosystem.

Just as it is clear that single factors control the sizes of some populations in some communities, it is also clear that the regulation of other populations cannot be explained so simply. One cannot explain the relative abundance of algae in lakes on the basis of phosphate concentration alone. Phosphorus may be the most significant limiting factor for most algal species, but there are other significant factors as well. One significant study by David Tilman demonstrated how 2 different species of diatom (small single-celled algae) have different growth responses to phosphorus and silica. Both species responded positively to increases in phosphorus availability. But their ability to coexist depended jointly on phosphorus and silica availabilities.[10] Zinc may be a limiting factor in some plants, but the requirement for zinc is reduced if the plant grows in the shade rather than in sunlight. Some organisms are able to substitute one element for another, which would be limiting if the substitute were lacking. Even in modern agriculture, fertilization is not designed to raise the concentration of "the" limiting factor, but rather to raise the level of all nutrients closer to their optimal levels.

Biogeochemical Cycles

The patterns of cycling matter in the biosphere involves not only metabolism by living organisms, but also a series of strictly abiotic chemical reactions. Understanding the cycle of a single element requires a feeling for a process that depends jointly on the biology of all organisms that utilize the element, its geological availability, and its organic and inorganic chemistry. Understanding the cycling of biologically important elements is truly an interdisciplinary endeavor: It is with good reason that their cycling through living organisms and back and forth between the community, water, sediments, and the atmosphere is termed *biogeochemical cycles*.

The movement of an element through a community can be viewed using the food chain model: This constitutes the organic phase of the biogeochemical cycle (Figure 3.5). The only difference between the organic phase of a biogeochemical cycle and the food chain as presented in Figure 2.8 is that the material is not lost to the ecosystem when a chemical is broken down, as in respiration. Rather, it enters the inorganic phase.

The inorganic phase describes the cycling back and forth between the

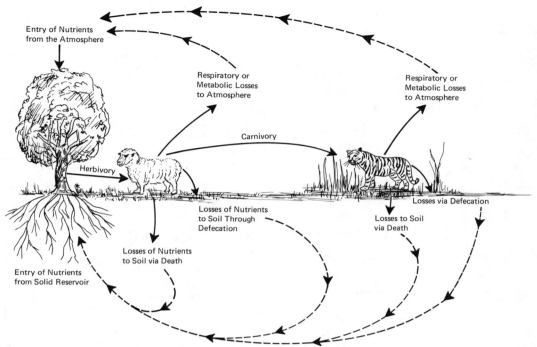

Figure 3.5
Organic phase of a biogeochemical cycle in a terrestrial environment.

reservoirs of the ecosystem shown at the beginning of this book (Figure 3.6). All biologically important elements can be found in all four reservoirs. The key to the inorganic phase is what controls how elements are made available to the organic phase. This is straightforward for some elements and in some ecosystems; it is very complex in others.

All biogeochemical cycles for all ecologically significant elements have both an organic and an inorganic phase. Both are critical: How efficiently the material moves through the organic phase determines how much is available to living organisms, but the major reservoirs for all metabolically important elements are external to the food chain. Furthermore, flow in the inorganic phases tends to be slower than in the organic phase.

Carbon Cycle

The biogeochemical cycle with the greatest resemblance to energy flow through the ecosystem is the carbon cycle (Figure 3.7). This is not surprising, because energy in living things is stored as fixed carbon.[11] Indeed, it makes very little difference when tracing movement through the community whether one is looking at chemical energy or carbon. Virtually all carbon enters the community in the form of carbon dioxide, which passes through the leaves or other photosynthetic parts of autotrophs, where it is converted to sugar by photosynthesis. It is combined with other materials and built into more complex compounds as it contributes to plant biomass and as it passes through the food chain. Some carbon is released as carbon dioxide when respiration takes place. It reenters the atmosphere, where it can be reused by plants.

This is the basic carbon cycle; it is simple and complete. But not all carbon

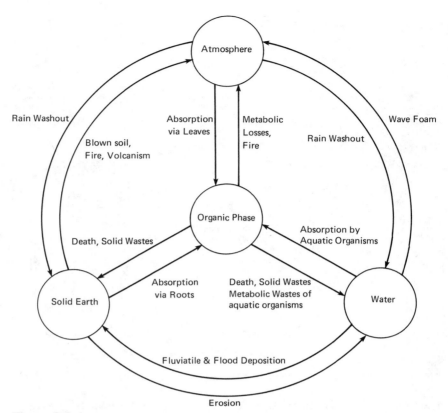

Figure 3.6
Generalized biogeochemical cycle, showing interchanges among the various reservoirs.

is respired. Some is fermented, and some is stored in solid form. The volatile carbon compounds that enter the atmosphere after anaerobic metabolism, such as methane and more complex molecules, are readily oxidized inorganically to carbon dioxide in the atmosphere. Carbon is stored in sediments as elemental carbon (e.g., coal), as refractory organic materials (e.g., the dark colored humic materials in soils or black shale), or as carbonate (e.g., limestone). Just as geological deposition acts to store materials, erosion can uncover them, and inorganic chemical weathering of rock can mobilize the carbon contained there. Some carbon is released as carbon dioxide from volcanoes and from similar examples of intense geological activity. In recent years people have greatly increased the rate at which carbon passes from sedimentary form to carbon dioxide. Burning fossil fuels recycles sedimentary carbon much faster than does natural weathering.

Some carbon, especially in the sea, is found not as organically fixed carbon, but rather as carbonate ($CO_3^=$), especially calcium carbonate ($CaCO_3$). This material is used for shells by animals such as clams, oysters, some protozoa, and some algae. Carbon dioxide reacts with water to form carbonate in the following three-step reaction:

$$CO_2 + H_2O \longleftrightarrow H_2CO_3 \longleftrightarrow H^+ + HCO_3^- \longleftrightarrow 2H^+ + CO_3^=$$

| carbon dioxide | water | carbonic acid | | bicarbonate ion | | | carbonate ion |

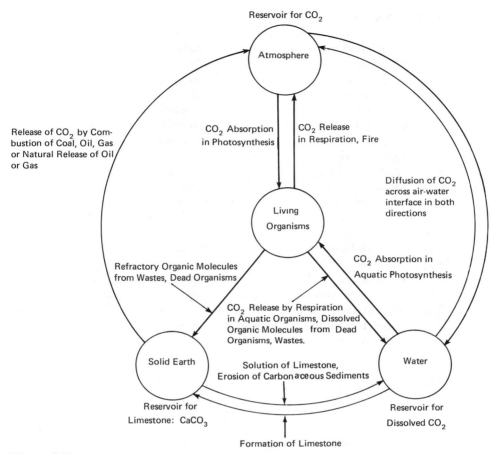

Figure 3.7
The carbon cycle.

The balance among these reactions, determining the precise amount of each of these constituents in the water, depends on the acid content (pH) of the water. Organisms such as clams can combine bicarbonate or carbonate with calcium dissolved in the water to produce calcium carbonate. After the death of the animal, this calcium carbonate may either dissolve or remain in sedimentary form. Reefs and other deposits of biogenic calcium carbonate can go through the rock-forming process to form limestone, a rock type that is rather common in the geological record.

The carbon cycle is typical in many ways of biogeochemical cycles. The organic and inorganic phases are closely intertwined, and the element can move rapidly from one to the other. Also, cycling can take place entirely within one phase or the other. Substantial amounts of carbon are found in the sedimentary and aqueous reservoirs. Carbon dioxide is uncommon in the atmosphere (0.03%), but it is so available to green plants that it can cycle very readily between the atmosphere and the community. The multiplicity of paths along which carbon can flow is typical of biogeochemical cycles in general. This leads to a well-buffered system that can ensure an adequate supply of carbon.

Figure 3.8
Structural formula
of L-Cisteine, a
simple amino acid
containing an
amino group
(shaded) and a sulf-
hydryl group
(hatched).

Nitrogen Cycle

Nitrogen is one of the most critical nutrient materials required by all living things. It is found in all *amino acids*, which are the building blocks for proteins. These organic acids are characterized by an amino group ($-NH_2$), which is a key part of the "peptide bond" that gives proteins their structure and many of their properties (Figure 3.8).

By far the largest reservoir in the biogeochemical cycle of nitrogen is the atmosphere (Figure 3.9). Unlike the atmospheric form of carbon (CO_2), however, nitrogen in the atmosphere is overwhelmingly molecular nitrogen (N_2), which is fairly inert chemically and can be utilized by only a few organisms. The more biologically significant reservoir is the aqueous reservoir, where nitrogen is dissolved in soil water or in the water of aquatic ecosystems in the form of nitrate (NO_3^-) or ammonia (NH_3 or NH_4^+). Dissolved fixed nitrogen enters the food chain through the roots of vascular plants and through the cell walls of nonvascular plants. It is incorporated by the plants into organic molecules such as amino acids and proteins, pigments, nucleic acids, and vitamins. These are passed through the food chain just like any other food substance. Excesses are excreted: Ammonia and related compounds are major components of animal urine. Nitrogen can also leave the food chain directly when burned by forest and grassland fire. In this case nitrogen is released into the atmosphere either as nitrogen gas or as one of the nitrogen oxides.

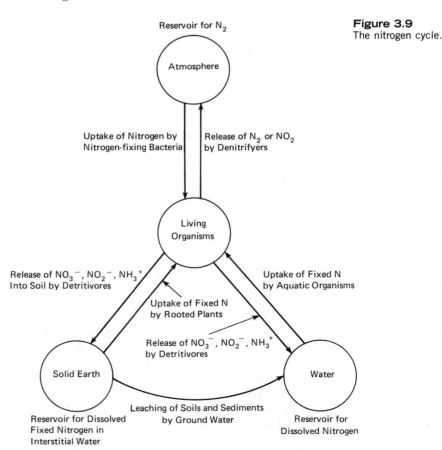

Figure 3.9
The nitrogen cycle.

Nitrogen is recycled through the detritus food chain. As nitrogenous wastes and carrion are degraded by detritus organisms, nitrogen is converted to the amino form (if it is not in this form already). The amino group ($-NH_2$) is liberated from the organic molecule to form ammonia (NH_3); this process is called *deamination*. Certain bacteria, most notably of the genus *Nitrosomonas*, can oxidize ammonia to nitrite by the reaction:

$$2NH_3 + 3O_2 \longrightarrow 2NO_2^- + 2H_2O + 2H$$

This reaction takes place in the soil, in lake or sea water, or in sediments. It is possible wherever ammonia is being released and oxygen is present. It is easiest to view this as a simple reaction occurring within the detritus food chain, although it is not strictly a part of the breakdown of organic carbon. As fast as nitrite is produced, other bacteria of the genus *Nitrobacter* can oxidize the nitrite to form nitrate:

$$2NO_2^- + O_2 \longrightarrow 2NO_3^-$$

The formation of nitrate from reduced forms of nitrogen is termed *nitrification*. The reactions involved release energy, so that the nitrifying bacteria are chemosynthetic autotrophs that can extract the energy for their metabolism from the chemical energy of the reduced nitrogen molecule, just as green plants can extract energy from electromagnetic radiation. Nitrate can then be taken up by plants at the beginning of the food chain. Thus the organic phase of the nitrogen cycle is a complete cycle.

Under certain circumstances, nitrate is either not produced in the nitrogen cycle or it is degraded before it can be absorbed by plants. Degradation of nitrate is termed *denitrification*, and it may be significant when oxygen concentration is low. Denitrifying bacteria can extract oxygen from the nitrate ion to break the nitrate down to nitrite, ammonia, or molecular nitrogen:

$$C_6H_{12}O_6 + 12NO_3^- \longrightarrow 12NO_2^- + 6CO_2 + 6H_2O$$

$$C_6H_{12}O_6 + 8NO_2^- \longrightarrow 4N_2 + 2CO_2 + 4CO_3^= + 6H_2O$$

$$C_6H_{12}O_6 + 3NO_3^- \longrightarrow 3NH_3 + 6CO_2 + 3OH^-$$

If denitrification is significant in an ecosystem, nitrite is transitory and does not last long before being degraded into ammonia or molecular nitrogen.

The fact that denitrification can transform fixed nitrogen into molecular nitrogen would eventually lead to a removal of all fixed nitrogen from living systems, were there no mechanism for fixing molecular nitrogen into a form usable by living organisms. But these mechanisms do exist, both inorganically in the atmosphere and biologically. First, lightning can oxidize nitrogen to form nitrogen oxides, which can then react with water to form nitrate:

$$N_2 + O_2 \longrightarrow 2NO \text{ (lightning)}$$

$$2NO + O_2 \longrightarrow 2NO_2 \text{ (inorganic chemical reaction)}$$

$$3NO_2 + H_2O \longrightarrow 2H^+ + 2NO_3^- + NO \text{ (reaction of } NO_2 \text{ with water)}$$

This is a slow process, but it produces nitrate at an approximate rate of 35mg/m²/year.

More important is biological fixation, whereby certain bacteria, fungi, and blue-green algae extract molecular nitrogen from the atmosphere or from aqueous solution and combine it with hydrogen to form ammonia. Some of this ammonia is excreted by the nitrogen-fixing organism, and thus becomes directly available to other autotrophs. Approximately 140 to 700 mg/m²/yr (up to 20,000 mg/m²/yr in fertile areas) enter the biosphere through this route.

Nitrogen-fixing organisms may be free-living, either in the soil or in water, and they produce vast quantities of fixed nitrogen. Other bacteria (notably *Rhizobium*) live in nodules produced on the roots of certain symbiotic vascular plants. These vascular plants include legumes (for example, clover, alfalfa, and soybeans) and some species of pine and alder. Several hundred species of plants can serve as hosts to nitrogen-fixing bacteria. When these plants grow in an area, the nitrogen content of the soil can rise appreciably. Indeed, the fertility of certain high-altitude mountain lakes can be determined by the species of alder surrounding them, depending on whether or not the trees contain symbiotic nitrogen-fixing bacteria.

The sedimentary phase of the nitrogen cycle is fairly simple. Nitrogen that is not reused by living organisms can be permanently incorporated into sediments by normal sedimentary processes. However, sedimentary nitrogen is either nitrate or ammonia. These are exceedingly soluble, and do not become bound onto soil particles as easily as other substances such as phosphate. They are likely to dissolve in water percolating through the soil and get carried into streams or into the groundwater system beyond the reach of autotrophs. The high solubility of nitrogen in soils has led to some serious pollution problems in areas where nitrogenous fertilizers are spread on soils and nitrogen is leached into nearby water supplies.

Sulfur Cycle

In sharp contrast to the carbon and nitrogen cycles is the sulfur cycle (Figure 3.10). Sulfur is not abundant in the atmosphere, but its geochemistry in the aqueous and sedimentary reservoirs is significant, and it parallels that of many other elements. Like nitrogen, sulfur appears in nature in several different forms. It is absorbed as sulfate ($SO_4^=$) through plant roots, where it is incorporated into certain organic molecules, such as some amino acids and proteins. It passes through the grazing food chain as other elements, with excesses being excreted in the feces. It can be introduced into the atmosphere through forest and grassland fires, in which case it is oxidized to sulfur dioxide.

The form of sulfur most commonly found in living tissues is in the sulfhydryl group ($-SH$), as part of an organic molecule (Figure 3.8). The sulfhydryl group is separated from the rest of the molecule by detritus bacteria, in the form of hydrogen sulfide (H_2S). This is a normal part of the degradation of proteins. In an aerobic environment, the hydrogen sulfide is oxidized to sulfate by bacteria specially adapted to perform this conversion. The sulfate produced can then be reused by autotrophs.

In anaerobic environments, such as the bottom of certain lakes, sulfide cannot be oxidized, since there is no free oxygen. However, there are photosynthetic bacteria which often inhabit such environments that can use infra-

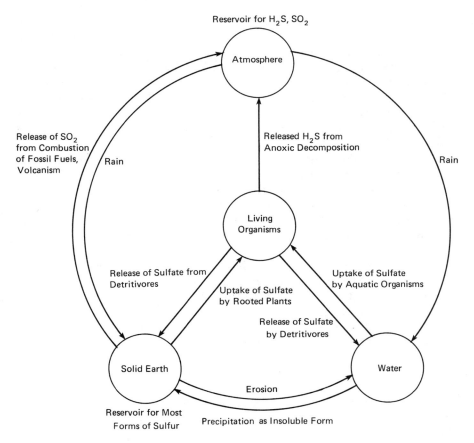

Figure 3.10
The sulfur cycle.

Reservoir for H₂S, SO₂

Atmosphere

Release of SO₂ from Combustion of Fossil Fuels, Volcanism

Rain

Released H₂S from Anoxic Decomposition

Rain

Living Organisms

Release of Sulfate from Detritivores

Uptake of Sulfate by Rooted Plants

Uptake of Sulfate by Aquatic Organisms

Release of Sulfate by Detritivores

Solid Earth

Water

Erosion

Reservoir for Most Forms of Sulfur

Precipitation as Insoluble Form

red radiation to manufacture carbohydrates and to oxidize sulfide either to elemental sulfur or to sulfate.

$$6CO_2 + 12H_2S + h\upsilon \longrightarrow C_6H_{12}O_6 + 6H_2O + 12S$$

$$6CO_2 + 12H_2O + 3H_2S + h\upsilon \longrightarrow C_6H_{12}O_6 + 6H_2O + 3SO_4^= + 6H^+$$

Elemental sulfur can also be utilized by other bacteria to form sulfate. If oxygen is present, the reaction is quite rapid. Elemental sulfur can also be oxidized to sulfate by certain bacteria under anaerobic conditions if nitrate is present. None of these bacterial reactions is unidirectional; there are conditions when other bacteria can reverse them.

These bacterial reactions provide a multitude of pathways for converting sulfur back and forth from its environmentally most significant form (sulfate) to forms that are stored readily in sediments (sulfur and sulfide). Sulfur is readily removed from the organic phase of the biogeochemical cycle, and the presence of dissolved iron in the ecosystem makes it even more insoluble. Even sulfate ions tend to be relatively insoluble, especially when they combine with certain common metals such as calcium. These metals may cause sulfate to form an insoluble mineral in the soil or sediment.

Removal of sulfur from the organic phase of the sulfur cycle can be counteracted in several ways by gains from the inorganic phase. Erosion of sulfate-rich rocks such as gypsum (CaSO₄ · 2H₂O) leads to gradual solution of the sulfate. Sulfur in other sulfide-rich rocks such as pyrite-bearing coal and shales can oxidize when the rocks are exposed to the air. The pyrite combines with free oxygen and water to form dissolved sulfuric acid. This form of sulfate is highly accessible to living organisms. But if a great deal of material is exposed (as with a coal mine), the pH of the acid water can be so low as to kill everything in streams affected by it. Oxidation of pyrite to sulfate can be completely inorganic, or it can be bacterial. In most instances, about 80% of the sulfate produced by pyrite oxidation is bacterial, and the remainder is inorganic.[12]

Sulfur is not abundant in the atmosphere anywhere, but the atmospheric phase of the sulfur cycle is significant, especially in industrial countries. Any organic sulfur that is burned, any hydrogen sulfide that escapes into the atmosphere, and any sedimentary sulfide that is released into the atmosphere (such as by the combustion of coal) is spontaneously oxidized by atmospheric oxygen to sulfur dioxide (SO_2). Some plants can utilize sulfur dioxide directly as it is obtained through their leaves. In addition, the soil itself may be able to absorb sulfur dioxide from the atmosphere. It has been estimated that 147 g/m^2 of sulfur dioxide is absorbed by soils in crops in Indiana over and above the sulfate that falls out in rain.[13]

Sulfur dioxide can also react with water to form sulfite ($SO_3^=$), which falls out in precipitation. Finally, and most important, sulfur dioxide can react photochemically with oxygen to form sulfur trioxide:

$$2SO_2 + O_2 + hv \longrightarrow 2SO_3$$

This is a relatively slow reaction, and its rate depends on such factors as the time of day and availability of metal catalysts in the atmosphere. About 1% of the atmospheric loading of SO_2 is converted to SO_3 per hour, on the average, but the rate appears to vary from about 0.5% to 5%. SO_3 can then combine with water to form sulfuric acid (H_2SO_4), which falls out in precipitation.

A significant amount of sulfate is added to ecosystems from industrial sources. The sulfate concentration in rain falling on areas uncontaminated by urban air pollution averages about 1.9 parts per million and the rain has a pH about 5.6. Rainfall in eastern North America and Northern Europe now has an average pH closer to 4.0. This represents an acid content about 40 times what would be expected. Most of the acidity in these areas is derived from sulfur in coal.[14] Episodes of much higher acidity have also been measured. One measurement in Pennsylvania showed a pH of 2.32, a 150-fold increase in acid content from the expected level,[15] and on in Wheeling, West Virginia was 7 times more acidic than that, with pH of 1.5.[16]

Phosphorus Cycle

The biogeochemical cycles of most biologically important elements other than those already described do not show much storage in the atmospheric reservoir. An example is the critically important element phosphorus (Figure 3.11). This element is found naturally in the environment as phosphate (PO_4^{---}, or one of its analogs, $HPO_4^=$ or $H_2PO_4^-$). It may occur as soluble inorganic ions, as part of a soluble organic molecule, as part of an insoluble organic or inorganic molecule, or as the constituent of a mineral grain as found in a rock or sediment.

The ultimate source of phosphate in the ecosystem (indeed of all nutrients

Figure 3.11
The phosphorus cycle.

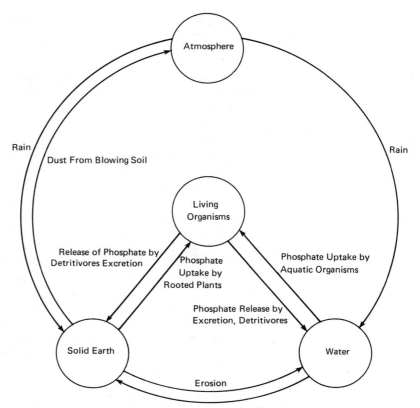

other than those abundant in the atmosphere) is crystalline rocks. As these erode and weather, phosphate is made available to living organisms, generally as ionic phosphate. Plants absorb it through their roots or across their cell walls and incorporate it into living tissues. Its significance to living things stems from its critical role in the molecules which control energy transfer within the cell. Phosphate is passed along the grazing food chain in the same fashion as nitrogen and sulfur, with excesses excreted in the feces. An extreme example of fecal phosphate (and nitrogen) is the tremendous guano deposits built up by birds on the desert west coast of South America. The Inca empire used guano as their primary fertilizer to feed the empire. Before cheap chemical fertilizers became common, the guano deposits were a major source of fertilizer for the world. Before active mining of the guano deposits began to decimate the beds in the last half of the 19th century, they were mountains 50 m high which had taken 2500 years to accumulate.[17]

Inorganic ionic phosphate is liberated into the abiotic environment as large organic molecules containing phosphate are degraded in the detritus food chain. It can immediately be taken up by autotrophs in this form, or it can be incorporated into a particle of sediment or soil. The organic phase of the phosphorus cycle is thus very simple. The complexities of the phosphorus cycle are in the inorganic phase, specifically in sediments. Inorganic phosphate that is not immediately used by living organisms can become bound onto and released from sediments. The amount of phosphate dissolved in the water percolating through a typical soil is very low compared to the amount found in the sediment. Relative to sediment-bound nutrient, it is much lower than, for example, dissolved nitrate.[18]

The reason for this is that phosphate is fairly insoluble at best, and it can react chemically with certain particles in the soil or sediment. Some of these reactions produce insoluble compounds that bind phosphate so tightly that it cannot be utilized by plants. For instance, phosphate reacts readily with aluminum ions in an aluminum-rich acid soil to form insoluble aluminum phosphate. Similar reactions take place with ions and with certain compounds containing calcium, iron, and manganese. Plants cannot utilize the phosphate bound up in the insoluble crystalline compounds formed by these reactions. How much phosphate is actually removed from the available nutrient pool in these ways depends on several factors, including the pH and the concentrations of various ions in the interstitial water in the sediment, the organisms found there, and the types of minerals and amount of organic matter that make up the sediment.

These simple inorganic chemical reactions are the most familiar to students, but they are overshadowed in their importance by different kinds of chemistry in the soil. Phosphate can be incorporated into the crystal structure of a class of minerals called *clay minerals*, which are abundant in most sediments and soils. This is a very complex process, and it is not fully understood. The clays are very reactive in several ways, and they form actively in most soils. The process involves strictly inorganic crystal growth, and it can lead to phosphate incorporation in sediments that is so tenacious that it becomes unavailable to living organisms.

Phosphorus can also be bound onto the *surface* of clay minerals and organic particles, through the process of *ion exchange*. This is especially important in soils, and it is the mechanism for retaining many nutrients in terrestrial environments in a form that is available to living things. The chemical bonds which allow surface adsorption on clays and organic particles are much weaker than those involved in inorganic reactions or clay crystal growth. They are strong enough to allow the nutrients to stay in place, even when the interstitial water that they would otherwise be dissolved in percolates rapidly through the soil. They are weak enough that plants can readily release them.

One could write volumes about the chemistry of phosphate and sediment particles, and this discussion has just begun to scratch the surface. It should be clear that although the organic phase of the cycle is straightforward, the inorganic phase is not. The availability of phosphate to living organisms is closely related to the rate of cycling through these two phases. And the way in which the phosphate cycle regulates the availability of phosphate to living organisms is what makes it so critical.

The availability of phosphorus to living things depends less on the absolute amount in the ecosystem than on the rate at which it is recycled. Especially in fresh-water lakes, the mobile phosphate pool is very small, and it may be absorbed by one organism as quickly as it is excreted by another. As long as organisms can meet their phosphorus requirement, they can grow. These needs can be met as long as phosphate cycles quickly. However, if the cycling rate slows down for any reason, phosphorus may be removed from solution. F. H. Rigler followed the rate of phosphate utilization using radioactive ^{32}P.[19] He showed that phosphate flow through the organic phase is strongly dependent on organic demand for phosphate. The length of time it

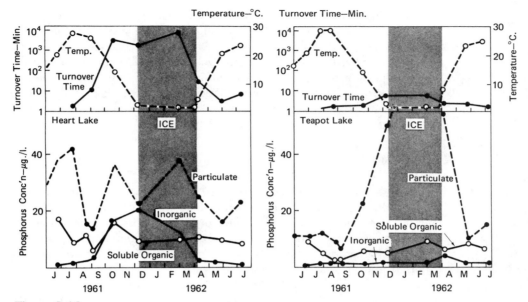

Figure 3.12
Relationships between temperature, turnover time of inorganic phosphorus, and concentrations of three forms of phosphorus as a function of time in two Ontario lakes. [Redrawn, with permission, from F. H. Rigler, "The Phosphorus Fraction and Turnover Times of Inorganic Phosphorus in Different Types of Lakes," *Limnology and Oceanography* 9:511–518. Copyright © 1964 by the American Society of Limnology and Oceanography, Inc.]

took for phosphate to proceed once through the organic cycle (*turnover time*) was exceedingly short during the summer, when demand for phosphorus was high, but it lengthened during the winter, when demand for phosphate was much lower (Figure 3.12). The average turnover time was on the order of 10 minutes in these fresh-water lakes. Studies of salt-water communities, have shown similar pattern, with summer turnover time much shorter than winter turnover time; the average was about 10 hours.[20]

It is much more difficult to measure the rates of flow through the sedimentary phase. G. E. Hutchinson suggests that the sedimentary turnover time in fresh-water ecosystems is likely to be measured in weeks rather than hours, and studies of terrestrial ecosystems indicate that turnover time may range up to 200 years.[21] If this is so, any change in the amount of phosphates released from the sediments is likely to be very sluggish. This hypothesis was tested in the Experimental Lakes Area in northwestern Ontario, Canada. Phosphorus was added to a lake during 1971 and 1972. Each year, the biota responded to this fertilization by producing tremendous *algal blooms*, in which algal populations expanded markedly to nuisance proportions. No phosphorus was added in 1973, and no algal bloom occurred that year. The phosphorus was still in the lake, but it had been scavenged by sedimentary chemistry from the biologically available pool.[22]

The phosphorus cycle is less straightforward than the cycles discussed previously. Like the other cycles, a multitude of pathways for nutrient movement exist, but they tend to be pathways to removal of the material from the biologically available pool, and they are inorganic rather than organic. As a result, the portion of the cycle that is under organic control assumes a critical importance in controlling the amount of phosphate available for living things. The net input of phosphate from outside the ecosystem through groundwater or surface water and the net rate at which it is incorporated into sediments are also extremely important. The sluggishness of the sedimentary phase of the phosphorus cycle and the fact that increased demand for phosphate by a biological community is met by increasing the rate of cycling through the organic phase rather than by releasing phosphate from the sedimentary phase is a major reason why phosphate is often the most critical nutrient in an ecosystem (see Chapter 7).

Other Nutrient Cycles

Biogeochemical cycles exist for all nutrients used by living organisms, as well as for some materials that are not nutrients. Most of the cycles are fairly simple, with the basic features shown in Figures 3.5 and 3.6 and with specific characteristics that depend on the inorganic chemistry and biochemistry of the material. The nutrient budget to a given ecosystem depends not only on these characteristics but also on the amount imported into the area or exported from it by wind and moving water.

Water is clearly the most competent medium for transporting nutrients from one place to another. It is tempting to ascribe the lion's share of import and export to moving water, so that the amount of nutrient available depends on the water budget. But it is becoming increasingly clear that the wind is an important medium for nutrient transport. For example, a substantial portion of the phosphorus loading in Lake Superior is from atmospheric dust carried in by the wind. Some elements are like phosphorus in that they are highly insoluble and tend to remain in the soil, being lost only when the soil itself is eroded. It is generally necessary to consider the specific characteristics of an ecosystem in working out the biogeochemical cycles of nutrients within it, including the biochemical requirements of the organisms in the community and the geochemistry of the material under the range of environmental conditions found in the area.

Cycles of Air and Water in Time and Space

One characteristic of biogeochemical cycles that has not been emphasized in the preceding discussion is that they tend to be fairly localized. We tend to talk about the phosphorus cycle of a lake or the sulfur cycle of an agricultural region. Even for the carbon and nitrogen cycles, where the atmosphere is such an important reservoir, our concern in discussing the cycles tends to be at the individual-ecosystem level. The basic question is, how much of the particular material is available to support life within the ecosystem, and what factors modulate this availability over a growing season or, perhaps, a year?

Many critically important factors in living systems need also to be explicitly examined on a much larger or even global basis. As with biogeochemical cycles, the movement of matter (especially air and water) from one place to another establishes a signal that generates a response in the living community. The input of living organisms to these global cycles is typically very small. However, they are part of the feedback system that regulates ecosystems and makes them what they are.

Atmospheric Circulation

Air moves over a global scale, and its movement determines the distribution of energy in the atmosphere, as well as the earth's weather and climate. It also controls whether pollutants introduced into the atmosphere are mixed or stagnate in place. Global circulation patterns arise because of the earth's rotation and because the atmosphere is heated from the bottom—like a kettle of water on a stove. As a result, air can circulate very freely over great heights under some conditions and be stagnant at other times. The basic pattern of global atmospheric circulation is shown in Figure 3.13. Winds tend to blow from the east and toward the equator in the equatorial regions, and they tend to be westerly in the temperate zones. Of course, these patterns represent the aggregate picture of prevailing wind directions; the actual pattern of wind movement at any particular time and place is much more complex.

Disruptions in atmospheric flow can cause an undulation in the boundary between two regions. The wave is often strong enough to spin off a mass of air from one zone and send it into the adjoining one. Extreme examples include hurricanes and typhoons, which spin off from tropical wind systems into the temperate zones, and massive cold fronts that enter the temperate zones from polar regions. These air masses may be either low-pressure or high-pressure bodies, depending on whether the atmospheric pressure within them is greater or less than the pressure of the surrounding air.

These pressure cells have at least two significant effects. Everybody who watches the evening weather on television is acquainted with their marked (and often very rapid) impact on weather. The passage of a cold front, for example, can cause temperatures to drop 10° to 20°C within a few hours in many parts of the temperate zone, as well as causing rain, snow, sleet, or hail. Less obvious, movement of pressure cells from one circulation zone into another is a key mechanism for transferring heat from the equator to higher latitudes. As a result, the tropics radiate less heat over a year's time than they receive in insolation, while the poles radiate much more.

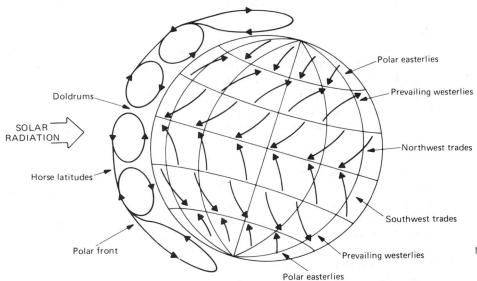

Figure 3.13
Basic patterns of global atmospheric circulation.

The global climate results from the way heat is absorbed by and transferred through the lower atmosphere. Both negative and positive feedback mechanisms exist to stabilize and to destabilize the climate as conditions change. Most of these changes now are due to human activity, but the geologic record shows that dramatic changes have occurred in the past. For example, the heat absorption characteristics of the atmosphere depend, in part, on its temperature. As the temperature rises, the amount of energy reradiated into space as infrared increases as well, but at a rate proportional to the fourth power of the absolute atmospheric temperature. This means that any tendency for the earth's temperature to increase or decrease would be counteracted by great decreases or increases in energy reradiated into space.

Water vapor absorbs infrared radiation very strongly, converting it to atmospheric heat. The more water vapor there is in the atmosphere, the more infrared radiation is absorbed, and the warmer the atmosphere will be. As the atmosphere gets warmer for any reason, more water vapor evaporates from the surface, and less condenses into clouds. Everything else being equal, increased absorption makes the atmosphere even warmer. This is a very strong positive feedback interaction between atmospheric temperature and the water vapor in the atmosphere. A similar effect is shown by carbon dioxide, which also absorbs infrared radiation strongly.

There is also a strong positive feedback interaction between temperature and snow. Snow has a much higher albedo than most materials at the earth's surface. A decline in temperature can increase the amount of snow cover and allow it to last longer. Everything else being equal, the increased albedo increases the amount of energy reflected into space and decreases the amount of energy absorbed at the surface. This leads to further cooling.

Cloudiness can either warm or cool the atmosphere by absorbing infrared radiation or by increasing the albedo. The actual effect of clouds depends on factors such as cloud height, amount of cloudiness, and the size and optical density of individual clouds. The atmosphere may respond in several different ways, and it is not at all clear what net effect can be expected.

The Hydrologic Cycle

Atmospheric circulation is the main carrier of heat from one part of the earth to another. But water has a critical role in heat storage and transfer. As it discharges this role, it also provides the fresh water which is so essential for all terrestrial ecosystems. The *hydrologic cycle* is the pattern of movement of water back and forth between the biosphere and the inorganic reservoirs. It is at the center of atmospheric heat transfer and biogeochemical cycles.

Like any other substance with a biogeochemical cycle, water enters into chemical reactions, both biochemical and inorganic. It moves along the food chain as the water which makes up much of organisms' bodies is ingested. Some animals can obtain all of their water needs from their food. In most cases, though, animals must also drink additional water, and virtually all animals excrete some water in their urine. Most of the water in the biosphere exists in an uncombined form throughout its cycle. Although water as a raw material of organic matter is as important as carbon dioxide, relatively little of the water found in living tissues is chemically bound: Living organisms are roughly 71% H_2O. Water is important for several reasons: It is the

Table 3.2
Distribution of Water
in the Earth's Surface
and Crust

Chemically Bound in Rocks: Does Not Cycle	
Crystalline Rocks	$250{,}000 \times 10^{17}$ kg.
Sedimentary Rocks	$2{,}100 \times 10^{17}$ kg.

Data from Hutchinson,
1957, Nace, 1967. Data
from Nace originally in
terms of volume; conversion factor 1 km.^{3} =
10^{12} kg.

Free Water: Moves Via Hydrologic Cycle	
Oceans	$13{,}200 \times 10^{17}$ kg.
Icecaps and Glaciers	292×10^{17} kg.
Ground Water to a Depth of 4000 m.	83.5×10^{17} kg.
Fresh-water Lakes	1.25×10^{17} kg.
Saline Lakes and Inland Seas	1.04×10^{17} kg.
Soil Moisture	0.67×10^{17} kg.
Atmospheric Water Vapor	0.13×10^{17} kg.
Rivers	0.013×10^{17} kg.

medium by which nutrients are introduced into plants; it is an important constituent of living tissue, either as liquid water or as part of essential organic molecules; it serves as a means of thermal regulation for both plants and animals; it is the medium by which sediments—a prime source of mineral nutrients—are removed from or added to local ecosystems; it covers the great majority of the earth's surface, and is the dominant feature of all aquatic ecosystems.

The hydrologic cycle is driven by solar energy and gravity. As was pointed out in Chapter 1, more than 80% of the total insolation that is not reradiated immediately goes to evaporate water. The atmospheric water vapor produced by this means can then condense around particles of dust in the atmosphere, which serve as *nucleation particles*, or particles that serve as the nuclei for water droplets. Eventually, the droplets are heavy enough to fall as precipitation under the influence of gravity. At its most basic, the hydrologic cycle can be viewed as an alternation of evaporation and precipitation, with the energy used to evaporate the water being dissipated as heat in the atmosphere when the water condenses. Water is not evenly distributed throughout the earth, as is shown in Table 3.2. Almost 95% of the earth's water is chemically bound into rocks and does not cycle. Of the remainder, about 97.3% is in the oceans, about 2.1% exists as ice in the polar caps and permanent glaciers, and the rest is fresh water: atmospheric water vapor, groundwater, soil water, and inland surface water.[23]

Water cycles rapidly between surface and atmosphere. The amount of water vapor in the atmosphere is sufficient, on the average, that it would cover the entire earth to a depth of 2.55 cm if it were in liquid form. But the average annual rainfall for the earth is about 81.1 cm (Furon, 1967), and it ranges up to 1,200 cm in some places. This means that the average turnover time for atmospheric water is about 11.4 days, or that the equivalent of all the water vapor in the entire atmosphere falls as precipitation and is re-evaporated more than 32 times per year.[24]

The geographic distribution of evaporation and rainfall is extremely uneven. Relatively more water precipitates onto land than evaporates from it (Table 3.3). If this were not the case, it would be difficult for terrestrial ecosystems to exist. However, some areas may get over 2500 cm of water per year, whereas some desert stations may go without water for over 14 years.[25] Still, more rain falls on the open ocean than on land, even taking into account the relative percentages of the earth's surface covered by land and sea.

The hydrologic cycle over the oceans is extremely simple (Figure 3.14).

Table 3.3
Water Balance of Oceans and Continents

Reprinted from R. L. Nace, "Are We Running Out of Water?" U.S. Geological Survey Circular 536.

Percentage of Earth's Surface			Precipitation or Evaporation (in kg. × 10¹⁷ per Year)
72	Ocean Surfaces	Evaporation from	3.50 (83.3%)
		Precipitation onto	3.20 (76.2%)
28	Land Surfaces	Evaporation from	0.70 (16.7%)
		Precipitation onto	1.00 (23.8%)
		Total Cycling Water	4.20 (100.0%)

But several things can happen to water that falls onto land: direct evaporation, transpiration through vascular plants, runoff, seepage into soil, and entry into groundwater systems, (Figure 3.15). Some of these processes are difficult to measure, because they are quite variable. We can lump these routes into three main categories: the rapidly cycling portion, or evapotranspiration, which includes the evaporation and transpiration, the less rapidly cycling water, or surface runoff, and the very slowly cycling groundwater. Water that seeps into the soil can end up in any one of these three categories. In the United States, estimates suggest that about 70% of the total rainfall goes to evapotranspiration, 27% to runoff, and 3% to groundwater.[26]

EVAPOTRANSPIRATION. *Evaporation* refers to water that evaporates directly from any surface other than a plant, such as a lake, soil surface, or animal skin. Its main effects in most cases are to moderate the temperature of a local area and to allow water vapor to enter the atmosphere. In some ecosystems, evaporation also leads to a concentration of salts in the water of a soil, which may be critical environmental factor.

Transpiration, on the other hand, is water that evaporates from the leaves of plants. From the viewpoint of gross water movement, a plant can be visualized as a hose leading from its roots to its leaves. Water enters the roots, is pulled through the plant body, and evaporates from the leaf surface. Very little of this water is removed for the manufacture of carbohydrates. However, the water that enters the roots of plants carries mineral nutrients. Thus, transpiration may appear in the balance sheet for water as a glorified form of evaporation, but its function in the ecosystem is much broader. It is the vehicle for minerals to enter the food chain via the roots of plants. It is thus a critical part of biogeochemical cycles in general.

The sum of evaporation and transpiration is termed *evapotranspiration*. The ratio of transpiration to total evapotranspiration varies from place to place, depending on the makeup of the community, temperature, relative humidity, wind conditions, etc. In non-irrigated field crops, transpiration

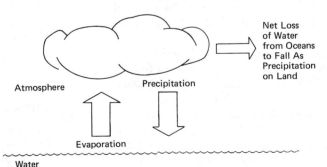

Atmosphere

Precipitation

Net Loss of Water from Oceans to Fall As Precipitation on Land

Evaporation

Water

Figure 3.14
The hydrologic cycle over the earth's oceans.

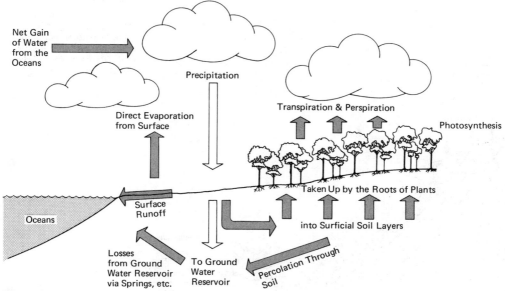

Figure 3.15
The hydrologic cycle over land.

typically accounts for slightly less than half the total. However, in some specialized environments, such as Japanese rice paddies, transpiration may account for 90% of the total evapotranspiration. Some communities are much more efficient in their use of water than others. A typical temperate field crop transpires 200–500 g of water for each gram of plant tissue produced. The figure may be double this in an arid region.[27]

SURFACE RUNOFF. If transpiration is the vehicle for nutrient uptake, then the gross movement of soluble and solid particles in the ecosystem is accomplished largely by runoff. Nutrients that have accumulated in sediments or soils can be eroded by streams and removed from the area. Ecosystems commonly show net gains or losses in specific materials. The most common agent for this in forests is soil erosion of soil by runoff. In one very intensely studied forest in New Hampshire, erosion from a small watershed into Hubbard Brook caused a net loss of several important nutrients, including calcium, magnesium, and sodium.[28] Obviously, the soil concentration of these nutrients would drop very low and survival of the community would be impaired if this continued for some time, and no mechanism for replenishing chemical nutrients existed.

The relationship between natural communities and running water to control nutrient losses is indicated by another study at the Hubbard Brook experimental forest. All vegetation was removed by clearcutting from a small watershed.[29] With the organic phase of the biogeochemical cycle artificially interrupted in this way, with the soil no longer being held by the tree roots, and with transpiration eliminated, there was more overall runoff, and the rate of nutrient loss increased greatly.

On the other hand, streams can carry nutrients eroded from upstream areas and deposit them in downstream ecosystems, thus raising their total fertility. This is one of the main reasons why low-lying wetlands such as swamps and estuaries, which are areas of accumulation for particulate mate-

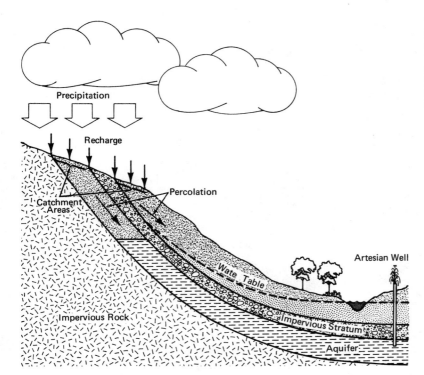

Figure 3.16
Diagram of the ground water system, showing water table, a porous rock aquifer, a spring, and an artesian well. [Redrawn from Ackermann, Colman, and Ogrosky, 1955.]

rial from other ecosystems, have among the highest productivity of any known ecosystems. Streams may also carry rock fragments which can be chemically altered through additional weathering until organisms can utilize the nutrient elements they contain. Finally, moving water erodes soil and allows weathering of the underlying rock to make their nutrients available to plants. The underlying rocks comprise the ultimate reservoir for all nutrients outside of nitrogen, oxygen, and hydrogen, the basic constituents of air and water. If erosion is very slow, the nutrients released through weathering can balance those lost to erosion, at least when added to those brought in with atmospheric dust. The role of moving water in the biogeochemical cycles of nutrients is as important in many ways as that of the organisms living there or as any of the organic or inorganic relationships we may think of as central to the biogeochemical cycle.

GROUNDWATER. Groundwater is the water that saturates either sediment or rock below the water table (Figure 3.16). In general, it is not tapped by plants for transpiration, and it is typically too deep to evaporate directly from the surface. It is an exceedingly important reservoir for water, accounting for more than 66 times the amount of water found in fresh-water lakes and streams. It is also an important economic resources in that it is a tremendous, but by no means infinite, source of water; indeed, it may be the only source of potable water in arid regions. Groundwater moves from one place to another under the influence of gravity. The area of net water movement from the surface into the groundwater system is termed a *catchment area*; areas where groundwater reaches the surface and runs off are termed *springs*. A rock body through which groundwater flows is an *aquifer*. A well drilled into an aquifer that has sufficient hydrostatic pressure to force water up into it is called an *artesian well*. The flow rate of ground water is exceedingly variable; it ranges from a few hundredths of a millimeter per day to as much as 5500 m/day, depending on the amount of water in the system and the characteristics of the aquifer.

Global Circulation in the Oceans

The hydrologic cycle is not the only significant circulation pattern involving water. The world's oceans are linked by a network of currents. The best known and best understood of these are at the surface (Figure 3.17), but many are at great depths. These currents determine the interchange between surficial and deep-water masses, as well as horizontal movement. Both the horizontal and vertical movements of ocean waters are significant, but for different reasons. A water mass retains its identity for great distances within a given current, such as the warm Gulf Stream, and the community associated with that water mass ranges far beyond what would otherwise have been its anticipated limits. In addition, the climate of terrestrial ecosystems is strongly affected by the nature of adjacent water masses. Ocean currents transport large quantities of heat from equatorial toward polar regions: Prevailing winds blow across the ocean and are heated or cooled by the waters they traverse. Thus Scandinavia, which is next to the warm North Atlantic Drift, has a relatively moderate climate despite the fact that it is as far north as Labrador.

The oceans' influence on the temperature of adjacent lands is also significant in the long term. Roughly three quarters of the world is covered with oceans. Their immense size and the high heat capacity of water gives them a powerful role in regulating the earth's temperature. The oceans can store or give up immense quantities of heat and hence confer a tremendous inertia onto incipient temperature change. This inertia is so great and so far beyond human control that it can delay warning signals of impending changes until they are inevitable. If, as many people fear, the buildup of carbon dioxide in the atmosphere is leading to an increase in heat absorption by the earth, so much of the heat would be stored in the deep ocean that global equilibrium would be delayed by decades. So much inertia would be provided by the oceans that people would be powerless to stave off the warming of the earth to reach that equilibrium. By the time it was clear what was happening, it would be too late.

The current system also sets the patterns of nutrient movement, and thereby controls the distribution of ocean productivity. Very little interchange occurs between surface and deep ocean waters over most of the sea. As a result, any nutrient that settles out of the upper layers does not return to the surface at the same location. The removal of required nutrients (especially nitrogen and phosphorus) from surficial waters has limited the productivity of most of the world's oceans to about the same level as a desert. At the same time, the deepest waters of the ocean are nutrient-rich. In certain places, notably at the west coasts of continents and in subpolar latitudes in both hemispheres, where deep currents rise to the surface (Figure 3.18) These *zones of upwelling*, have a high nutrient concentration that makes them sites of intense productivity, even though their waters are very cold (Figure 2.7). They constitute only about 0.1% of the ocean surface, but they yield almost 60% of the total worldwide marine fish catch. Nutrients brought

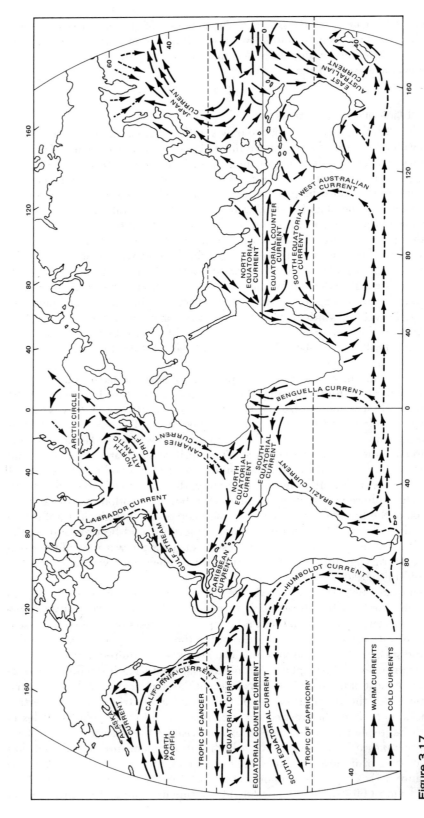

Figure 3.17
Surface currents of the world oceans.

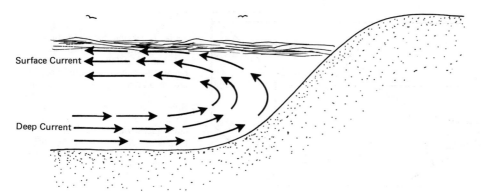

Figure 3.18
Diagram of a zone of upwelling adjacent to a continent.

to the surface in an upwelling zone are then cycled back to the rest of the ocean by means of surface currents.

The Geological Materials Cycle

In many ways, the ultimate but most sluggish and least complete cycle of all is the geological materials cycle (Figure 3.19), by which the solid materials of the earth are regenerated and made constantly available. The cycle is basically a balance between the leveling effects of erosion and the constructive effects of geological uplift. There would be no terrestrial ecosystems if there were no uplifts and if gravity and rainfall doomed the landscape to endless erosion. Any primeval mountains would long since have been carried by moving water to the sea, and the sediments deposited in the sea would have displaced enough water to inundate whatever land was left.

When nutrient materials are carried either in rivers or by the wind, they are deposited in basins of one sort or another. The ultimate depositional basin for all sediments is the sea. In certain areas of maximum sediment accumulation, termed *geosynclines*, sediments may build up for millions of years. These sediments are compressed by intense lateral forces generated by moving plates within the earth's crust. As a result, they are typically destabilized and upthrust into mountain ranges. Classic examples are the Rocky Mountains and the Alps, which were once the final depositional site for more than 2 billion km^3 of sediment, representing the organic matter and mineral nutrients lost from land areas over a span of some 400 million years. The mountains formed from these sediments are now supplying habitat and mineral nutrients for a new generation of terrestrial ecosystems. Mountain building may be accompanied by volcanism, as in the Cascades and Andes of the Americas or the island arc of Japan. Surficial materials are taken deeper into the earth where two of the great plates making up the crust come together, and new materials emerge at crustal plate boundaries in the midoceanic ridges. Thus, from the perspective of geologic time, even the most apparently permanent losses to ecosystems may be recycled.

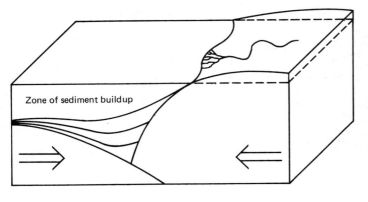

Zone of sediment buildup

Figure 3.19
Driving force for geological materials cycle. A continental plate (right) moves leftward relative to the oceanic plate. Sediments build up between the plates, but pressures increase as plates continue to move. Ultimately, these pressures result in widespread melting and metamorphism of rock in the subsurface, coupled with uplift of mountain ranges and volcanism at the surface. This subjects the materials incorporated into the original sediments to a new round of erosion.

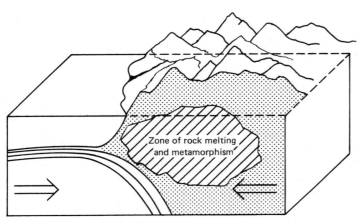

Zone of rock melting and metamorphism

Notes

[1] McMillan, C., 1959. Salt tolerance within a *Typha* population. *Am. Jour. Botany*, **46**, 521–526.

[2] Phleger, C. F., 1971. Effect of salinity on growth of a salt marsh grass. *Ecology*, **52**, 908–911.

[3] Paine, R. T., 1974. Intertidal community structure: Experimental studies on the relationship between a dominant competitor and its principal predator. *Oecologia*, **15**, 93–120.

[4] Ratcliffe, D. A., 1967. Decrease in eggshell weight in certain birds of prey. *Nature*, **215**, 208–210; Cade, T. J., Lincer, J. L., White, C. M., Roseneau, D. G., and Swartz, L. G., 1971. DDE residues and eggshell changes in Alaskan falcons and hawks. *Science*, **172**, 955–957; Peakall, D. B., 1974. DDE: its presence in peregrine eggs in 1948. *Science*, **183**, 673–674.

[5] McMillan, C., 1960. Ecotypes and community function. *Am. Nat.*, **94**, 246–255; McNaughton, S. J., 1966. Ecotype function in the *Typha* community type. *Ecol. Monog.*, **36**, 297–325.

[6] Gibbons, J. W., and Sharitz, R. R., 1974. Thermal alteration of aquatic ecosystems. *Am. Sci.*, **62**, 660–670.

[7] Oldfield, J. E., 1972. Selenium deficiency in soils and its effect on animal health, *Geol. Soc. Am. Bull.*, **83**, 173–180.

[8] Schindler, D. W., 1974. Eutrophication and recovery in experimental lakes: Implications for lake management. *Science*, **184**, 897–899.

[9] Blackman, F. F., 1905. Optima and limiting factors. *Annals of Botany*, **19**, 281–295.

[10] Titman, D., 1976. Ecological competition between algae: Experimental confirmation of resource-based competition theory. *Science*, **192**, 463–465; Tilman, D., 1980. Resources: a graphical-mechanistic approach to competition and predation. *Am. Nat.*, **116**, 362–393.

[11] Garrels, R. M., Lerman, A., and McKenzie, F. T., 1976. Controls of atmospheric O_2 and CO_2: Past, present, and future. *Am. Sci.*, **64**, 306–315.

[12] Kuznetzov, S. I., Ivanov, M. V., and Lyalikova, N. N., 1963. *Introduction to Geological Microbiology*. New York: McGraw-Hill Book Company, Inc.

[13] Burtrumson, B. R., 1950. Sulfur studies of Indiana soils and crops. *Soil Science*, **70**, 27–41.

[14] Likens, G. E., Wright, R. F., Galloway, J. N., and Butler, T. J., 1979. Acid rain. *Sci. Am.*, **241**(4), 43–51; Shinn, J. H. and Lynn, S., 1979. Do man-made sources affect the sulfur cycle of northeastern states? *Env. Sci. Tech.*, **13**, 1062–1067.

[15] Commonwealth of Pennsylvania, 1979. Position paper on the interstate transport of air pollution, October 24, 1979. Harrisburg, Pa.: Department of Environmental Resources.

[16] Alexander, G., 1982. Running on Empty? *Nat. Wildlife*, **20**(1), 42–47.

[17] Paulik, G. J., 1971. Anchovies, birds, and fishermen in the Peru current, in Murdoch, W. W., ed., *Environment*. Sunderland, Mass.: Sinauer Associates, 156–185.

[18] Thomas, G. W., 1970. Soil and climatic factors which affect nutrient mobility, in Engelstad, O. P. (ed.) *Nutrient Mobility in Soils: Accumulation and Losses*. Soil Science Society of America, Special Publication 4, pp. 1–20.

[19] Rigler, F. H., 1964. The phosphorus fractions and the turnover times of inorganic phosphorus in different types of lakes. *Limnol. and Oceanog.*, **9**, 511–518.

[20] Pomeroy, L. R., 1960. Residence time of dissolved phosphate in natural waters. *Science*, **131**, 1731–1732.

[21] Reiners, W. A., and Reiners, N. M., 1970. Energy and nutrient dynamics of forest floors in three Minnesota forests. *Jour. Ecol.*, **58**, 497–520.

[22] Schindler, D. W., 1974. Eutrophication and recovery in experimental lakes: implications for lake management. *Science*, **184**, 897–899.

[23] Nace, R. L., 1967. Are we running out of water? *U. S. Geol. Surv. Circ.*, **536**, 1–7.

[24] Furon, R., 1967. *The Problem of Water: A World Study*. New York: American Elsevier Publishing Co., Inc.

[25] Garstka, W. U., 1978. *Water Resources and the National Welfare*. Fort Collins, Colo.: Water Resources Publications.

[26] Nace, R. L., 1965. Status of the International Hydrological Decade. *Jour. Amer. Water Works Assn.*, **57**, 819–823.

[27] Chang, J-H., 1968. *Climate and Agriculture: An Ecological Survey*. Chicago: Aldine Publishing Co.

[28] Likens, G. E., Bormann, F. H., Johnson, N. M. and Pierce, R. S., 1967. The calcium, magnesium, potassium, and sodium budgets for a small forested ecosystem. *Ecology*, 48, 772–784; Bormann and Likens, 1979. *Pattern and Process in a Forested Ecosystem*. New York: Springer-Verlag.

[29] Bormann, F. H., Likens, G. E., Fisher, D. W., and Pierce, R. S., 1968. Nutrient loss accelerated by clear cutting of a forest ecosystem. *Science*, **159**, 882–884.

Epilogue

The physical signals that link the elements of ecosystems are many and varied. Because of the global circulation patterns of the winds and ocean, the ecosystems of the world are linked to a significant degree. But the patterns of interaction on the local level are especially significant. Whether we are talking about energy or nutrients, species pass materials to one another, and their survival depends on their meeting their needs for nutrient and energy intake and satisfactory excretion (and subsequent breakdown) of waste products. The patterns that control the movement of materials in ecosystems are not chaotic. The notion of the food chain gives a first indication of the differences in roles of different populations. These roles will be refined as we discuss the ecological niche in Chapter 6. Ecosystems are regulated by an intense linkage of many different kinds of signals. They are similar in many ways to human systems, with their information feedbacks. But the concrete signals of energy and materials form a critical part of the linkages in ecosystems.

There are profound differences in the characteristics of different ecosystems. In some areas, such as mountainous regions, the soils and the nutrients they contain are constantly being eroded, so that the organic phase of nutrient cycles must be very efficient, as must the community's mechanisms for making available the nutrients contained in the rocks underlying the soil. On the other hand, ecosystems such as estuaries may be depositional areas for both energy and nutrients from other areas, and they may be exceedingly productive as a result. Ecosystems, then, are not equal. All have characteristics of their own, which must be taken into account if people are to exist as elements of productive ecosystems. The natural flow of energy and nutrients through ecosystems is fundamental and necessary to their continuation. An alteration that in one area might be trivial and easily balanced by existing natural feedback mechanisms might disrupt the total balance in others.

Further Reading

Ackermann, W. C., Colman, E. A., and Ogrosky, H. O., 1955. From ocean to sky to land to ocean, in U.S. Department of Agriculture Yearbook of Agriculture for 1955, *Water*, 41–51.

Brady, N. C., 1974. *The Nature and Properties of Soils*, 8th ed. New York: The Macmillan Company.

Christiensen, N. L., 1973. Fire and the nitrogen cycle in California chaparral. *Science*, **181**, 66–67.

Fortescue, J. A. C., 1979. *Environmental Geochemistry: A Holistic Approach*. New York: Springer-Verlag.

Geiger, R., 1965. *The Climate Near the Ground*, 4th ed. Cambridge, Mass.: Harvard University Press.

Hutchinson, G. E., 1957. *A Treatise on Limnology*. Vol. 1, Geography, Physics, and Chemistry. New York: John Wiley & Sons, Inc.

Lieth, H., 1964. Versuch einer Kartographischen Darstelling der Pflanzendecke auf der Erde. *Geographisches Taschenbuch 1964–1965*, 72–80.

Lieth, H., and Whittaker, R. H., 1975. *Primary Productivity of the Biosphere*. New York: Springer-Verlag.

Mertz, W., 1981. The essential trace elements. *Science*, **213**, 1332–1338.

Odum, E. P., 1963. *Ecology*. New York: Holt, Rinehart, and Winston, Inc.

Odum, E. P., 1971. *Fundamentals of Ecology*, 3rd. ed. Philadelphia: W. B. Saunders Company.

Odum, H. T., 1957. Trophic structure and productivity of Silver Springs, Florida. *Ecol. Monographs*, **27**, 55–112.

Pomeroy, L. R., 1970. The Strategy of Mineral Cycling. *Ann. Rev. Ecol. Syst.*, **1**, 171–190.

Reichle, D. E., Franklin, J. F., and Goodall, D. W., eds., 1975. *Productivity of World Ecosystems*. Washington, D. C.: National Academy of Sciences.

Rice, E. L., and Pancholy, S. K., 1972. Inhibition of nitrification by climax ecosystems. *Am. Jour. Bot.*, **59**, 1033–1040.

Rumney, G. R., 1968. *Climatology and the World's Climates*. New York: Macmillan Publishing Co.

Rumney, G. R., 1970. *The Geosystem: Dynamic Integration of Land, Sea, and Air*. Dubuque, Ia.: Wm. C. Brown Company, Publishers.

Slobodkin, L. B., 1962. Energy in animal ecology. *Adv. Ecol. Research*, **1**, 69–101.

Teal, J. M., 1957. Community metabolism in a temperate cold spring. *Ecol. Monog.*, **27**, 283–302.

Teal, J. M., 1962. Energy flow in the salt marsh ecosystem of Georgia. *Ecology*, **43**, 614–624.

U. S. Air Force Cambridge Research Laboratories, 1965. *Handbook of Geophysics and Space Environments*. Bradford, Mass.: U.S. Air Force Cambridge Research Laboratories.

Van Valen, L., 1971. The history and stability of atmospheric oxygen. *Science*, **171**, 439–443.

Westlake, D. F., 1963. Comparisons of plant productivity. *Boil. Rev.*, **38**, 385–425.

Whittaker, R. H., and Likens, G. E., eds. The primary production of the biosphere. *Hum. Ecol.*, **1**, 301–369.

Organisms in Ecosystems

III

LIVING THINGS do more than make ecosystems somehow different from other natural systems. True, there would still be a hydrologic cycle if there were no communities, and nutrients would still cycle, at least to a degree, by means of the geological materials cycle. But the order we perceive in natural ecosystems would not exist. We could describe patterns, but they would be classifications of chaos. One of the sources of excitement in ecology is that we can find patterns that apply to different ecosystems in different times and places. We can see changes in the structure of ecosystems which clearly stem from the activities of organisms, but which involve a dynamic interaction between living and nonliving phases of the ecosystem, with the organisms changing the environment, and the environment changing the community.

The predictability of many ecosystem features and the prevalence of responses which appear to be driven by stabilizing negative feedback can be misleading at times. It is sometimes easy to think of ecosystems as "superorganisms" or as goal-seeking systems. We are accustomed to orderly systems designed by people to carry out a specific function. It is easy to forget that the force that has forged the order and consistency we see in ecosystems is natural selection acting over a very long period of time. The fossil record, despite its well-known tendency to provide only cryptic information, supplies abundant evidence that evolution has shaped ecosystems, albeit slowly, and that the order we see in nature is, in fact, natural.

This book is based on the premise that ecosystems are systems; that is, they have elements and signals linking those elements. The elements are the physical things that can be pointed to and described: populations and abiotic variables (e.g., bodies of water, landforms, the energy input from the sun). The signals are things that move among the elements. They are more abstract than the elements: They can be described and measured, but they need not be concrete (e.g., flow of nutrients in biogeochemical cycle, scent marking of territories by mountain lions, visual perception of a mouse by a circling hawk).

This book is based on the premise that ecosystems are cybernetic systems; that is, they are controlled by feedback within the system. Some element

"A" provides a stimulus that brings about a response from a different element "B." This response is itself a stimulus to "A" and causes it to alter its behavior. The time frame of the feedback loop may be as short as the time it takes for a hawk to capture a rabbit or as long as the time it takes for a forest to colonize a barren sand dune. In the first case, "A" is the rabbit, "B" is the hawk, and the behavior is a well understood response to visual stimuli. In the latter case, "A" is the biological community, and "B" is the soil. Plant growth on the dune provides material to improve the soil to a degree that new plants can enter the community. Community and soil develop together, until the forest is virtually indistinguishable from other area forests built on other foundations (see Chapter 6).

An ecosystem contains many species. Each has a different role, as well as different adaptations to discharge that role. One can study living things in their ecosystems from several different vantage points, but two are especially useful. The first deals with the individual species; the second concerns the means by which different species fit into a total community.

The population and the community are conceptually quite distinct. They have different properties and characteristics, and the properties of a population cannot be predicted by summing the properties of all the individuals in the population. In the same way, the community is much more than the sum of the populations of which it is composed. The notion that "the whole is greater than the sum of its parts" is universal in complex systems. The sum of parts is just that, as a pizza is the sum of the pieces into which the pizza is cut. But the whole includes not only the parts, but also the interactions of the parts. The properties of systems that stem from these interactions, and that cannot be perceived without taking an overview of the system, are the *emergent properties* of the ecosystem.

It is a little like the blind persons who were led to an elephant and asked to describe it. They came up with vastly different descriptions, depending on whether they felt the trunk, the legs, or the ears. They were all right, of course, but an elephant is more than all of these things—it is also an elephant. Communities are composed of populations, just as populations are made up of individuals. Communities constitute a part of a larger ecosystem. We need to understand all of these levels in order to understand the operation of natural ecosystems. We cannot comprehend the whole without understanding the parts, and none of the elements give a complete picture of the whole.

Populations

4

A POPULATION is the total assemblage of individuals of a given species found in any ecosystem under study. It is generally the most appropriate focus for studying species in their ecosystems. The properties of populations include things like abundance and population density, range, genetic variability, adaptations, resource needs and other demands on the environment, and factors that influence population growth, birth and death rates, and age distribution. These properties can be studied from a static viewpoint, concentrating on the population's characteristics at any one time, or from a dynamic viewpoint, emphasizing growth, changes, and regulation through time.

Regulation of Population Size and Density

One of the most obvious characteristics of any population is its size. This can be measured in several ways, including abundance (absolute numbers in the population), numerical density (number of individuals per unit area or volume), and biomass density (biomass per unit area or volume). All of these measures have different significances. Counting or weighing organisms in relation to the area or volume in which they are found allows the comparison of different sized ecosystems. It might seem significant, for example, if data showed that one lake contained 10 million fish, while another contained only 1 million. However, if the first lake was 10 times as large as the second, the density of the two lakes would be the same. In the same way, 5,000 animals are more than 2 animals, but the numerical difference does not seem so significant if the first are rats and the second rhinoceroses.

Some measures of population density recognize that only a portion of an ecosystem may be suitable for colonization by a particular population. For example, the proportion of a mature spruce forest that is suitable for infestation by the spruce budworm (a voracious consumer of evergreen trees) is large. The proportion in a mixed forest containing only a few trees suitable for the budworm would be much smaller. Comparing the intensities of population growth in two forests, one composed almost exclusively of spruce and one mixed, would require a measure of population density based on the amount of space suitable for the species rather than the total space in the ecosystem. Measures of this sort are often termed *ecological density*.

Some populations maintain a constant density over a period of time, while others fluctuate widely. These variations reflect the responses of the population to signals from all facets of the environment. Some of these signals represent abiotic factors, such as limiting factors and the availability of critical nutrients; others represent biotic factors, which are the interactions of different populations and instinctive control mechanisms that are internal to the population itself.

The *regulation* of population density describes the patterns of interaction between the controlling factors and the population in question. What kinds of stimuli are provided by the other populations or by abiotic factors in the environment? How does the population under study respond to these stimuli? One of the best indicators of the efficiency of this regulation is the population's *stability*: its uniformity over time. Note that there is no value judgment in how tightly a population is regulated. "Stable" is not "good," and "unstable" is not "bad." Stability simply indicates the effectiveness of the *homeostatic* (variation-resisting) mechanisms acting on the population.

Population Growth in Fluctuating Environments

The basic resource requirements of animals and plants were discussed in Chapter 3. No organism can survive unless it can meet these. But simply being able to meet resource needs does not ensure a population's stability or even its survival.

Population growth in many organisms does not appear to be regulated by their interactions with other species. They provide an extreme case of growth as a response to stimuli from the abiotic environment. There may be no evidence at all of feedback that would alter growth as a response to the influences of the population's growth. These species are as close to unregulated as one can find in an ecosystem; they behave as the system shown in Figure 1.4*a*. But even here, growth and decline are responses to signals generated by the larger environment.

Most such organisms respond to the random variations of climatic variables such as temperature and rainfall. The population increases steadily during favorable conditions and falls to almost zero during unfavorable conditions. Figure 4.1 shows an example. The density of blue-green algae in Lake Erie is normally quite low. Every August, however, conditions are such that they increase dramatically to an "algal bloom" which covers the surface with a musty green mass and confers a terrible taste to drinking water drawn from the Lake. Similar fluctuations in abiotic variables, notably seasonality and weather, have been suggested as the prime factor regulating the density of many populations.[1]

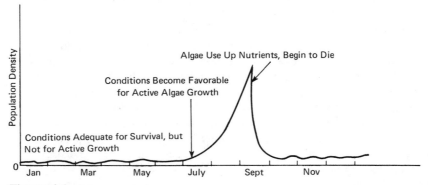

Figure 4.1
Diagrammatic representation of algae growth in Lake Erie throughout the year.

Living organisms have a tremendous influence on the changes and regulation of each other's population densities. However, the mechanism for these influences is very different from that of the abiotic factors. The latter represent the population's resource needs or constitute the landscape within which its members carry out their lives. It is meaningful to describe these needs in abstract terms such as tolerance range and optimum. The biotic factors that regulate populations are more active. The *interspecific* factors are direct pairwise interactions between species, and the *intraspecific* factors represent interactions of individuals within a single species.

One of the things that facilitates understanding the abiotic factors is that the interactions of ecosystem elements involving nutrients and similar things can be visualized in terms of a concrete signal linking the elements. Chemical or physical stimuli generated by one population (or the sun or a soil) and received by another population (or a body of water or a soil) can be seen as that, measured, and understood. The feedback loops based on these signals tend to be slow and cumbersome, but they tend also to be fairly straightforward. Biological factors are not this simple. Some are based on concrete signals, others on information about the abstract representations of things that are interpreted by the receiver.

Interspecific Factors

The great variety of interactions involving living things has spawned a complicated classification based on the general roles of the populations in the interaction. The mechanism for the interaction within a particular class, however, may depend on the adaptations of the organisms involved. For example, the behavior of higher animals is often an active response to interpreted information. For example, two animals can compete for a nesting site by fighting, bluffing, or threatening each other. The signals they give each other may be visual, aural, or tactile. They respond to those signals with motion and behavior. Plants, on the other hand, also compete for resources, such as phosphorus in the soil. They have neither muscles nor nervous system, so they cannot consciously receive information from other species or respond actively to them. But the signal exists, and there is a response just the same. Just as the animal less able to compete for nesting space must find another area, the tree that is less efficient at removing phosphorus from the soil may die.

The results of interspecific interactions may be a relative increase, parity, or decrease in the population size compared with what would have been the case in the absence of the interaction. The interaction may be *obligatory* (i.e., necessary for survival) to an individual of one or both of the species, or it may not. The interaction may act throughout the entire range of one or both of the species, or it may be restricted to a portion of their range. The basic classes of interspecific interactions are summarized in Table 4.1.

NEUTRALISM. Neutralism is the most common type of interspecific interaction. Neither population directly affects the other. What interactions occur are subtle and indirect. The mere presence of the two species should not directly affect the population level of either. An example of neutralism would be robins and squirrels living in a forest. Neither serves as food for the other,

Table 4.1
Types of interactions between individuals of two species, called "A" and "B"

	Effect of Interaction on Individuals of "A"		
Effect of Interaction on Individuals of "B"	Positive	Neutral	Negative
Positive	Mutualism	Commensalism	Predation Parasitism
Neutral		Neutralism	Amensalism
Negative			Competition

neither has any direct interaction except that both may inhabit the same tree, yet both are clearly part of the forest community.

COMPETITION. Competition refers to the interaction in which two individuals or species vie for limited amounts of food, water, nesting space, or other resources. Its result is that the sum of the equilibrium densities of the species is less than the sum of the densities they would have had in the absence of competition, and at least one effect of the competition interaction is a depression in both species' net reproductive capacity in some way.[2] Interspecific competition is not a simple interaction, and it may have many different outcomes, depending on the organisms involved.

As a general rule, when two species are in direct competition with each other for the same set of resources, the one that is more efficient at obtaining those resources and translating that efficiency into more effective reproduction will survive, and the other will die out. If the requirements of the two species are slightly different, both may remain in the community. This phenomenon is known as the *competitive exclusion principle*.[3]

Competitive exclusion is much like the food chain. It is a useful model for understanding certain key aspects of competition, but it is too simple to explain the way real ecosystems operate. A classic example of competitive exclusion, however, will outline the principles of competition effectively.[4] The protozoan genus *Paramecium* has many species with similar resource

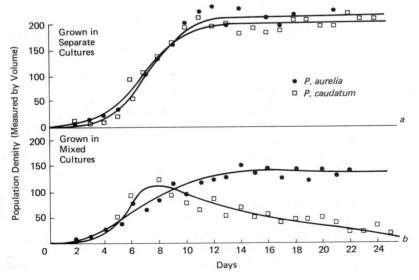

Grown in Separate Cultures

Grown in Mixed Cultures

- ● *P. aurelia*
- □ *P. caudatum*

Population Density (Measured by Volume)

Days

Figure 4.2
Growth of two species of *Paramecium* occupying the same niche: (*a*) species are each grown in pure culture; (*b*) species are in competition with one another. [Modified, with permission, from G. F. Gause, *The Struggle for Existence.* Copyright © 1934 by The Williams and Wilkins Co., Baltimore.]

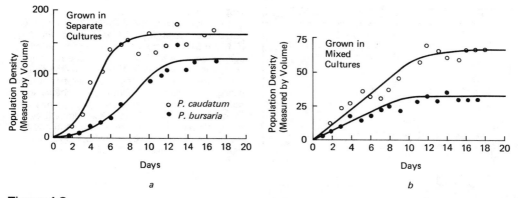

Figure 4.3
Growth of two species of *Paramecium* occupying slightly different niches: (*a*) species are grown in pure culture; (*b*) species are in competition with one another. [From data of Gause. Redrawn, with permission, from F. S. Bodenheimer, *Problems of Animal Ecology*. Copyright © 1938 by The Clarendon Press, Oxford.]

requirements. *P. aurelia* or *P. caudatum* both reach stable population densities when they are grown separately (Figure 4.2). They survive together for a while, when grown together in mixed culture, but *P. caudatum* is eventually driven into local extinction, and the level of *P. aurelia* reaches the original equilibrium level. On the other hand, if *P. caudatum* and *P. bursaria* are grown together in mixed culture, both survive and reach a stable population size (Figure 4.3). This is possible because their space requirements are somewhat different, even though their feeding requirements are similar.

The problem with competitive exclusion is that it holds that populations cannot coexist in a community if their resource requirements are identical. In fact, the resource requirements of real populations in real communities are never identical. The slight differences in species' requirements can allow several species to coexist, even if they all utilize the same resources to some degree.

For example, two diatom species, *Asterionella formosa* and *Cyclotella meneghiniana*, require both phosphorus and silica to survive. The amount of both of these nutrients can be shown diagrammatically by a gradient of silicate to phosphorus ratio, or Si/P (Figure 4.4). Phosphorus limits growth in both species when Si/P is high, and silica is the limiting factor when Si/P is low. There is a point at which the two nutrients are present in the correct proportion. This point occurs at a much higher Si/P level for *Asterionella* than for *Cyclotella*. One possible reason is that *Asterionella* is more efficient

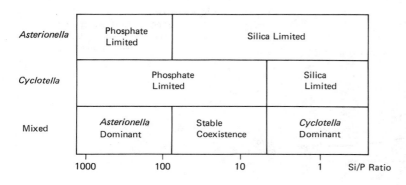

Figure 4.4
Schematic responses of two genera of diatoms to limitation by silica and phosphorus. The upper two bars indicate which nutrient limits the algae at different ratios of the two nutrients. The boundary is the level at which both nutrients are equally limiting. The lower bar indicates the result of competition at different nutrient ratios. [Data from Tilman, 1977.]

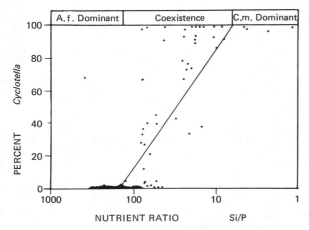

Figure 4.5
The relative abundance of *Cyclotella* (C.m.) (compared to the total of *Cyclotella* and *Asterionella* [A.f.]) in samples from Lake Michigan plotted against ambient silicate to phosphate ratios in the same samples. Notation at the top of the figure shows predicted outcomes of competition; the line shows the expected proportion of *Cyclotella* in the region of coexistence. Redrawn, with permission from Tilman, D. 1977. Resource competition between planktonic algae: an experimental and theoretical approach. [Ecology 58: 338–348. Copyright © Duke University Press.]

at taking up phosphate at low concentrations than *Cyclotella*, while *Cyclotella* is more efficient at taking up silica at low concentrations than *Asterionella*. As a result, *Asterionella* can be expected to be the dominant species under low-phosphate conditions, and *Cyclotella* can be expected to predominate under low-silica conditions. In between, however, they can coexist. Figure 4.5 shows the percentage of each species in a natural Si/P gradient in Lake Michigan. The results are precisely as expected.[5]

We can generalize these results into a simple graphical model that shows the results of competition between two species competing for two resources. Figure 4.6 shows the gradient of the two resources (R_1 and R_2), and two lines representing species A and B. In both cases, the species can survive when R_1 and R_2 are above this line, and they cannot at lower resource levels. When resource levels are below both lines, neither species can survive; when resource levels are below one and above the other, one species survives and the other does not. This is not competition; it is simple survival based on resource availability. The shaded area in the figure above both lines is the area in which competition takes place.

The outcome of competition depends on which population is *more* efficient at consuming the resource that limits the *other*. Figure 4.7 shows the four possibilities. Assume, for the moment, that each species is limited by a resource and much as the *Asterionella* and *Cyclotella* discussed in the previous paragraph. The species that is less effective at consuming the resource that limits it operates at a distinct disadvantage. If one species has an "advantage" and the other has a "disadvantage," then the former will exclude the

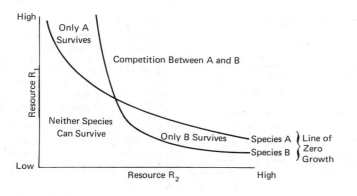

Figure 4.6
Schema for occurrence of competition between two species in an environment containing two potentially limiting resources.

Figure 4.7

Outcomes of interspecific competition based on relative ability of species to acquire limiting resources.

		Relative Efficiency of Species B at Acquiring Resource Limiting Species A	
		B Better Than A	A Better Than B
Relative Efficiency of Species A at Acquiring Resource Limiting Species B.	A Better Than B	Both species Coexist	A Outcompetes B
	B Better Than A	B Outcompetes A	Outcome Depends on Initial Conditions

latter. They can coexist in the ecosystem when both operate at a "disadvantage," and either can exclude the other when both operate at an "advantage," depending on the initial conditions.[6]

The classical view of competition is for resources: The competitive advantage lies with the species better able to *exploit* the resources presented by the environment. The reason for the advantage is that, as with the diatoms discussed earlier, the more efficient competitor reduces the availability of the resource to the less efficient one. Competition can also operate by direct *interference* of one species by the other. One of the early demonstrations of interference competition was in a set of experiments showing the role of temperature and relative humidity on the outcome of competition between two species of flour beetles.[7]

An example from an actual ecosystem is the interaction between two species of crayfish in the Little Sioux River of Minnesota and Iowa.[8] *Orconectes virilis* and *O. immunis* are widely distributed east of the Rocky Mountains in North America, with very similar ranges. However, they are seldom found together, and their resource requirements are very similar (both are generalized scavengers that eat anything available, living or dead), which suggests competitive exclusion. In fact, *O. virilis* does very poorly in lakes and ponds subject to drying, and it cannot withstand the low oxygen concentrations found in stagnant water. So the exclusion of *O. virilis* from the stagnant water of the upper Little Sioux river is not competitive; it is resource-based.

Both species prefer rock as a substrate. However, crevices in rocks that provide needed hiding places for crayfish are limited, and *O. virilis* is dominant over *O. immunis* in obtaining them. The mechanism is active aggression by *O. virilis* individuals against *O. immunis*. As a result, *O. immunis* is excluded from areas of the Little Sioux whose bottom is rocky, and *O. immunis* is relegated to muddy areas in portions of the Little Sioux that have mixed mud, gravel, and rock substrates.

Some competitive situations involve both exploitation and interference. In the back-reef zone on the east side of Discovery Bay, Jamaica, two sea urchins (*Diadema antillarum* and *Echinometra viridis*) compete for food with each other, and with the three-spot damselfish (*Eupomacentris planifrons*). All are herbivores, feeding on algae associated with staghorn coral (*Acropora cervicornis*). All three inhibit the others' population growth: When individual sea urchins of either species are removed, the other expands markedly. When three-spot damselfish are removed, the densities of both sea urchin species increases. *Diadema* is a large, active urchin, and it is capable of reaching new food sources quickly when they appear. *Echinometra* is smaller and slower, but it can defend a food supply against even the larger

Diadema. The two urchins, then, exploit the same resource using much the same basic adaptations, but with different strategies.

The key to the competitive interaction is the damselfish, which actively interfere with the sea urchins. They defend explicit territories (see following, page 99), and will attack urchins (and other organisms) that enter their territories by biting their spines and attempting to carry them from the area. The damselfish attack *Diadema* preferentially, but they can also pick up an *Echinometra* readily if they have to. By concentrating on the more active urchin, they alter the balance between the two species. It is likely that they also provide a stabilizing influence on their population densities.[9]

MUTUALISM. Mutualism is an interaction that benefits the species involved. Obligate mutualism is also known as *symbiosis*. Mutualisms are extremely common in nature, their manifestations are often quite striking, and they have been among the most important evolutionary factors operating in ecosystems. The most widespread mutualisms are between animals and plants, to use the animals to improve the efficiency of plants' reproduction and to provide food for the animals in return.

Examples of mutualism are well known. Lichens are a colony consisting of algae and fungi which live closely together and supply the colony with resources that the other cannot. The alga produces food photosynthetically, and the fungus takes in water and minerals. Bacteria in the rumens of cattle and the zooflagellates in the guts of termites can digest cellulosic materials. As a result, the cattle and termites can consume grass and wood and have it digested. In both cases, the large animal is totally dependent on the single-celled organisms in its gut to digest the cellulose, and the bacteria or zooflagellates are totally dependent on their partner to provide them with food.

Mutualism can even exist between two species of higher animal. For example, water moccasins and large birds such as herons and ibises on several islands off the west coast of Florida have an unusual but mutually beneficial interaction. The birds nest in the lower branches of relatively unprotected trees, while the snakes congregate around the bases. This protects the birds from tree-climbing predators such as raccoons. In turn, the snakes feed, in part, on fish dropped by the birds and the occasional baby bird that falls out of the nest.[10]

The most striking types of mutualism are those involved with plant reproduction. Virtually all plants have a sexual stage in their life cycle, in which genes are recombined to increase the genetic variation of the population. In archaic plants such as ferns and their allies, the sexual stage was diminutive and required a moist environment. In seed plants, pollen is released from the male organ and carried by some medium (either an animal or the wind) to the female fruiting organ, where fertilization takes place. The seed grows, is released from the seedcase, and is dispersed into the environment. It germinates and produces a new plant.

Perhaps the most dramatic evolutionary development of vascular plants was the origin of angiosperms, or flowering plants. The overwhelming majority of plants alive today are flowering plants. All of the early flowering plants, and most modern forms, require an animal to carry the pollen to the stigma so that fertilization can occur. The entire purpose of showy flowers, sweet nectar, and similar attractants is to entice insects, birds, bats, or other pollinating animals to the flower to pick up pollen and carry it to another flower. The plant depends on the animal to complete its life cycle. The animal uses the plant as a source of food.

Some plant-pollinator mutualisms are relatively weak. Several animal species can pollinate the plant equally well, and a given pollinator visits several different species of plants. Other symbionts are totally dependent on their partners for survival as a species. For example, the yucca plant of the desert areas of northwestern Mexico is pollinated exclusively by the yucca moth, and reproduction of the yucca moth occurs exclusively in the yucca plant. Both reproduce at the same time. The adult female moth visits the plant, collects pollen, and places it in the stigma. Only she has the proper-shaped mouth parts to put it in place. She then lays her eggs in the ovary which she has just fertilized, and goes on to a different stigma. By placing the pollen on the stigma, she has allowed fertilization to take place and seeds to grow. Her offspring grow in the ovary, feeding on developing seeds and ovarian tissue which are growing because of her pollination. Because the moth caterpillars are relatively inefficient consumers, however, plenty of seeds are left to disperse when they hatch and leave the plant. The two species are totally dependent on each other.

There can be no question that the yucca and its moth have evolved together. But other pollination systems have resulted from different evolutionary stresses. For example, two understory herbs in Panama, *Costus alleni* and *C. laevis*, share the same pollinating bee, *Euglossa imperialis*. They also occupy the same habitats, flower at the same time, and have identical flowers and nectar secretion patterns. Both species are fed upon heavily by a weevil which destroys 30 to 60% of total flower production. It appears that the similarity of the flowers represents evolutionary convergence in response to herbivory by the weevil. This allows two species which are competitors in other respects to increase the regularity and rate of pollinator visits, thereby increasing their seed set.[11]

Seed dispersal is another focus of mutualism. For example, the seeds of many small forest plants are carried by ants. Each seed typically has an ant-attracting body (the elaiosome), which is full of high-quality ant food. The ants are attracted to the ripe seeds and carry them to their nests. The ants benefit by eating the elaiosomes, and the plants benefit from having their seeds "planted" in the nutrient-rich soil of the ant nest.[12] In addition, ant-dispersal of seeds greatly reduces seed losses to seed-eating rodents. In one observation in West Virginia, 24 to 49% of the seeds of *Asarum canadense* were lost to predation when ants were allowed to disperse the seeds. However, 70% were lost when ants were excluded.[13]

Other plants have their fruits dispersed by birds. Birds are attracted to the fruits and eat them. They pass through the gut and are defecated shortly afterward. In the meantime, the bird has flown from the site of the parent plant, and the seed is left in a supply of fertilizer, ready to germinate. Some plants in temperate North America even time their fruiting to allow maximum exposure to migrating frugivorous birds.[14] As if to underscore the significance of obligate mutualism to both partners, the tambalacoque tree (*Calvia major*) of the Indian Ocean island of Mauritius has a very heavy seed, which may well have been dispersed and prepared for germination by the Dodo bird, which has been extinct since 1681. The tree used to be a common timber-producing tree whose wood was exported. Only 13 dying individuals remained in 1973, and their ages were estimated at over 300 years. If the connection between the Dodo and the tambalacoque is, in fact, valid, the tambalacoque's reproductive cycle was broken once the Dodo was driven into extinction by the Dutch.[15]

Another example of mutualism with considerable significance to human ecosystems is root infections. Bacteria or fungi can become established in the fleshy parts of the roots of vascular plants. They feed off the food produced photosynthetically by the plant, but they "pay their way" by improving the plant's ability to absorb nutrients, fixing nitrogen into a form that it can utilize, or detoxifying certain constituents of soils. *Mycorrhizae* are the most common of these and can be found in most species of vascular plants. These fungi improve the nutrient absorption characteristics of the plant.

Experiments comparing the growth and survival of plants inoculated with mycorrhizal fungi with those of uninoculated control plants demonstrate that the association commonly improves growth and may even be the difference between life and death under stress conditions, such as drying of the soil. Mycorrhizae may even be essential for the survival of most plants in poor tropical soils.[16] Nitrogen-fixing *Rhizobium* bacteria infect the roots of legumes and certain other plants and provide fixed nitrogen not only to their hosts, but also to the remainder of the community. Their role has already been discussed along with the nitrogen cycle on page 56. Nitrogen-fixing organisms other than *Rhizobium* include certain blue-green algae and some colonial relatives of the bacteria called actinomycetes.[17] Interestingly, the existence of mycorrhizae on legume roots increases the likelihood of *Rhizobium*-containing nodule formation. There are other root-based bacteria-vascular plant pairs. For instance, the bacterium *Beggiatoa* has been found associated with rice in anoxic rice paddy soils. The bacterium can detoxify hydrogen sulfide, which is toxic to the rice root, and the rice can detoxify peroxides that inhibit the bacterium.

COMMENSALISM AND AMENSALISM. Commensalism and amensalism are two types of interspecific interaction that, in many ways, bridge the three kinds of relationships discussed so far. Only one species seems to be affected. Commensalism bridges mutualism and neutralism; amensalism bridges competition and neutralism. The commensal species benefits from the interaction with its host, but the host is not affected. It is common in nature and is typified by the interaction between a tree and the birds that build their nests in the tree. The relationship is obligatory for the bird; it has essentially no effect on the tree. Other examples of commensalism could be cited in which the relationship was beneficial but not obligatory for the commensal species.

Amensal species are at least inhibited by the inhibitor, and they are sometimes excluded from an area altogether, but the inhibitor is not particularly affected by the amensal.[18] The most common examples of amensalism involve antibiotics secreted by plants which effectively keep other plants from growing near or under them. Black walnut (*Juglans nigra*) secretes the antibiotic juglone, which inhibits other trees, shrubs, grasses, and herbs to some degree; some plants, however (including black raspberry and Kentucky bluegrass) are tolerant of juglone and are often found beneath black walnuts. In the chaparral vegetation of California, certain species of shrubs, notably *Salvia leucophylla* (a mint) and *Artemisia californica* (sagebrush), produce volatile toxins that accumulate in the soil during the dry season and reduce

Figure 4.8
Amensalism between plant species in the California chaparral. Volatile toxins produced by *Salvia leucophylla*, the shrub at left of the photograph, prevent germination or growth of grasses in a zone about 2 m. wide and inhibit many grassland species in a zone extending an additional 3–8 m. The exclusion zone appears as a bare ring surrounding the shrubs (between A and B); the inhibition zone appears as an irregular ring of stunted individuals of relatively few species (between B and C). [Reprinted, with permission, from C. H. Muller. "The role of chemical inhibition (Allelopathy) in vegetational composition." *Bull. Torrey Bot. Club*, **93**, 332–351. Copyright 1966 by the Torrey Botanical Club. Photograph courtesy of Dr. C. H. Muller.]

germination or growth of grasses in a band up to 1 to 2 m wide surrounding them (Figure 4.8).[19] Certain species in mature ecosystems can produce substances that inhibit nitrification and, in essence, retard changes in community composition.[20]

PARASITISM AND PREDATION. Parasites and predators obtain food at the expense of their hosts and prey. These processes are basic to the entire grazing food chain above the autotroph level (there is no functional difference between a herbivore "preying on" a plant and a carnivore preying on another animal). Predators tend to be larger than their prey and catch them from without. A parasite is smaller than its host and consumes it ether from within (as with tapeworms or canine heartworms) or from without (as with ticks or body lice).

It is easy to believe that because the predator-prey interaction causes the consumption of prey individuals, it is somehow detrimental to the prey population. This belief has led to extensive "predator control" efforts in the name of wildlife conservation. But the coevolution of species within natural ecosystems has led to a dynamic balance between the populations in a community, so that the population sizes of predator and prey species are interregulated by feedback mechanisms that control the populations of both species (Figure 4.9). In a very oversimplified scenario, an increase in prey density above some reference level is equivalent (to the predator) to an increase in food supply. An increased nutrition base for the predator means

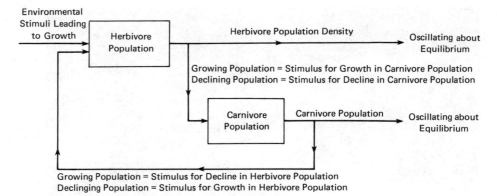

Growing Population = Stimulus for Decline in Herbivore Population
Declinging Population = Stimulus for Growth in Herbivore Population

Figure 4.9
Simplified diagrammatic representation of feedback interactions between a population of herbivores and a population of carnivores.

increased predator survival and consequently greater predation pressure on the prey. Greater predation pressure drives the prey density down. Conversely, a decrease in prey density leads to a decrease in predator food supply, which leads to lower predator survival, reduced predation pressure, and greater prey survival.

The oversimplifications in this scenario stem mainly from the fact that the responses of predators and prey to each others' densities is not instantaneous, and it may not even be very rapid, so that the population density of one species tends to lag behind the other. Also, predators and prey seldom constitute an absolute pair, so that the predator has only one prey species and the prey has only one predator. Real organisms typically eat, and are eaten by, many different species. Also, predators have to find their prey, and few ecosystems are homogeneous, featureless places where all prey individuals are equally easily caught. The results of an experimental predator-prey pair are a cycling rise in the prey, followed by a rise in the predator, followed by a fall in the prey, followed by a fall in the predator, and back through the cycle (Figure 4.10). The cycle keeps going until the predators locate and consume all of the prey. We shall return to this theme in Chapter 5.

Predator species can drive their prey into extinction. This does no particular harm to the predator if it preys on several other species at the same time; it simply switches its predation pressure. However, predators that depend

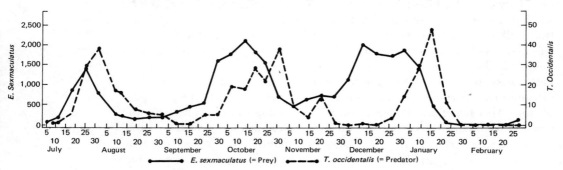

Figure 4.10
Interdependence of population size of two species of mites. *Eotetranychus sexmaculatus* serves as the food supply for *Typhlodromus occidentalis*. [Reprinted from C. B. Huffaker, "Experimental Studies on Predation: Dispersion Factors and Predator-Prey Oscillation," *Hilgardia*, **27** (1958) 343–383.]

on one or two prey species cannot afford to destroy them, because this would undercut its own resource base and the predator would die out as well. In such cases, populations that are not as effective at driving prey into local extinction have a selective advantage over those that are more efficient predators, because their genes are more abundant in the species at times of very low predator and prey density.

Some of the best illustrations of mutual population regulation by predation are insect pests of commercial crops. These are well documented, because of their economic importance, and examples are known from many crops and from around the world.[21] One of the best understood examples involves the cottony-cushion scale insect, *Icerya purchasi*, and the vedalia beetle, *Rodolia cardinalis*.[22] The scale insect was unintentionally imported from Australia into California about 1870. It is not an important pest in Australia, and it serves as the main food source for several insects, notably the vedalia beetle and a predaceous fly, *Cryptochaetum iceryae*. But these natural predators were not introduced into North America along with *Icerya*, and the scale insect found an abundant food supply in the citrus crop of southern California. It expanded rapidly through the citrus groves, unchecked by any natural predator. People feared that the insect would continue to expand until it had destroyed its own food supply—and the California citrus industry in the bargain; indeed, this was already a distinct possibility by 1886.

Cryptochaetum was deliberately introduced into California in 1886 and began at once to feed on the scale. But it could not keep up with the vast numbers of *Icerya* in southern California. At the end of 1888, *Rodolia* was introduced into the orange groves, and they consumed so many scale insects within a year that the cottony-cushion scale problem was brought under control. From 1890 until the mid-20th century, the scale and the vedalia beetle lived together in a dynamic equilibrium. The scale insect never amounted to more than a minor nuisance, but their numbers were sufficient to ensure a constant food supply for the vedalia beetle.

Soon after World War II, the use of pesticides (most notably DDT and organophosphorus types such as parathion and malathion) became widespread in the citrus-growing regions of southern California. The justifications for pesticide use were several, including not only the control of pests on noncitrus crops but also the final eradication of the cottony-cushion scale. Unfortunately, the vedalia beetle is much more sensitive to these pesticides than the scale insect. A primary result of pesticide spraying was the destruction of the vedalia population. This was followed by a dramatic rise in the density of *Icerya purchasi* to create an economically important problem for the first time in more than 50 years.

Herbivory can also be a significant element in population regulation, although plants tend to be regulated less through consumption than herbivores, everything else being equal.[23] This is especially important with plants that have invaded a community from outside and have few or no natural enemies in the new community. A classic example of this involves the prickly pear cactus, *Opuntia*, which was introduced into Australia from South America during the late 19th century. Because no Australian control mechanisms could check the spread of this cactus, it quickly expanded throughout millions of acres, excluding cattle and causing a substantial economic hardship (Figure 4.11a). *Cactoblastis cactorum*, a cactus-eating moth from Argentina, was introduced in 1925. By 1930, *Opuntia* was under control in Australia (Figure 4.11*b*,*c*), and it has been regulated effectively by the moth since that time.[24]

There are many other significant examples of the regulation of population density through predation. A population of moose (*Alces alces andersoni*) became established on Isle Royale, Michigan, in 1908.[25] The herd expanded to such a high density by 1930 that they seriously damaged forest and pond vegetation and experienced periodic die-offs because of inadequate winter food supply. A population of wolves became established on Isle Royale around 1948. Wolves and moose quickly formed a close predator-prey pair, with the wolf as the only predator of the moose and the moose constituting 75% of the wolves' food supply. Predation of moose by the wolves has resulted in a substantial drop in the moose population to a relatively stable equilibrium level. Regrowth of vegetation is now much greater than at high moose density, and die-offs of moose have ceased.

One could argue from these examples that predation was "bad" for the cottony-cushion scale in California and the cactus in Australia, since it resulted in a stable population density far below the maximum. But this is an anthropomorphic view, and it could as easily (and as anthropomorphically) be argued that predation was "good," since the population density after predation in both cases was more stable than it had been previously. Predation cannot be considered detrimental in any way to the moose population on Isle Royale. In fact, all of these animals have evolved within a regimen of population regulation in which predation plays a prominent role. Removal of predation pressure necessitates a more extreme regulatory mechanism, which is not normally a factor controlling the equilibrium level of the population. Population eruptions can overtax food supplies, leading to an extreme "boom and bust" cycle.

Normal predation does more than contribute to the stability of the population. Predators tend also to select the weakest prey individuals in the population. This removes less-fit genes in preference to more-fit genes. Consumption by animals also stimulates movement of materials through the food chain, and limited cropping of plants by herbivores stimulates further plant growth and reduces the senescence in plant populations.[26] All of these factors may contribute materially to the functioning and continuity of the population as a whole—and they may even be essential.

DEFENSE. Predation and parasitism are not one-way interactions. Animals (and even plants) under predator attack respond to that attack. The existence of predation means a strong selective advantage for the development of any effective defense mechanism on the part of actual or potential prey.

Many defense mechanisms are behavioral: Animals under attack do things differently from animals that are not under attack. For example, a flock of starlings is normally a relatively disorganized assemblage of birds. But when threatened by a hawk from above, they bunch tightly. The hawk can easily capture an individual bird on the wing, but it is likely to run into several others if it attacks the tightly bunched flock, and its body is too fragile to withstand high-speed collisions with several birds at once. Many herding animals defend themselves from attack by grouping together with the largest males in the front facing the attacker, and the young behind with the females.

Figure 4.11
Control of population size of
prickly pear cactus (*Opuntia* sp.)
at Chinchilla, Queensland, Australia by the moth *Cactoblastis
cactorum:* (*a*) dense growth of
Opuntia in belar (*Casuarina*)
scrub prior to the introduction of
Cactoblastis (October, 1926); (*b*)
the same area after the introduction of *Cactoblastis;* cacti largely
destroyed (October, 1929); (*c*)
the same area with trees cut and
a dense growth of Rhodes grass
(December, 1931). [Photographs
courtesy if W. H. Haseler, Department of Lands, Queensland.]

a

b

c

Cryptic coloration is another adaptation used by many animals to escape predation. They blend in with the background so that they cannot be seen. The flounder in the sand of the sea bottom, adjusting its color to the color of the sand, is a prime example. Many moths and butterflies are colored to resemble tree bark, lichens, or leaves.

Some of the most interesting adaptations are chemical. Many plants and animals produce chemical substances that are toxic, foul smelling, or nauseating, to reduce the probability of predation. The best known are probably skunks. Many animals producing defensive chemicals are, like the skunk, strikingly marked to stand out as a warning to potential predators. Monarch butterflies raised on milkweed plants retain some very toxic cardiac glycoside chemicals in their bodies, produced by the plants, and are very distasteful to birds. Their striking coloration is sufficient to avert predation from any bird that has ever captured a monarch in the past. Significantly, the monarch's coloration is such a good warning to predators that other palatable butterfly species, notably the viceroy, resemble the monarch and benefit from birds' avoidance of them. This phenomenon, by which one species resembles another to benefit from its warning coloration, is termed *mimicry*. It is widespread in the animal kingdom.

Other animals incorporate predators' organs of attack into their own bodies as defense mechanisms. Many marine polychaete worms and nudibranch snails graze on sea anemones, for example. Undigested stinging cells from the anemones end up in the polychaete and nudibranch skins, where they protect their new hosts from attack by other predators.

Plants themselves produce defense chemicals.[27] The allelopathic chemicals involved in amensalism (see page 90) are only one example. Buttercups (*Ranunculus sp.*) contain protoanemonin, an irritant that can be fatal to livestock under certain conditions. Cows do not eat them until other forage is gone. Nicotine is a powerful insecticide (and is commercially used as such to control some insects). It paralyzes aphids on tobacco plants. The cardiac glycosides in foxglove (*Digitalis purpurea*) can cause convulsive heart attacks in animals who eat them. A single leaf of oleander (*Nerium oleander*) is potentially fatal to humans. Tannins, which are widely found in plant bark, bind proteins into indigestible complexes, and they also inhibit fungal growth and virus transmission. Certain plants even produce powerful insect hormones that can prevent certain larval insects from molting into adulthood.

Plants also have physical means of defense. Spines such as those on desert plants (e.g., cacti) or semi-arid savanna plants (e.g., *Acacia*) are well known, but there are others. For example, the leaves of most plants are covered with tiny projections, or trichomes. These keep insect larvae or other herbivores away from the vulnerable portion of the leaf. Some trichomes are especially effective and can actually capture herbivores and prevent their escape.[28]

Intraspecific Biotic Factors

Not all biotic factors involve the interaction between species. Many population control mechanisms are intraspecific. They are never completely separate from interspecific and abiotic environmental factors, but they operate with these factors and may often be key elements in ensuring the stability of animal populations. They may be passive, as with intraspecific competition among plant individuals, or they can be active, based on the social organization of the population.

COMPETITION. Competition for scarce resources among individuals within a species is a feature of all populations. No ecosystem provides infinite resources to all of the individuals born to all species inhabiting it, and all populations are limited by their resource needs and their ability to acquire those resources.

Perhaps the clearest examples of intraspecific competition are in the plants, since plants do not mask competition with social mechanisms as do animals. At the minimum, they require nutrients, water, and access to light. One can demonstrate competition with respect to any of these resources, but two will suffice. Juvenile canopy trees in a mature forest typically grow very slowly in the understory, shaded by their conspecifics in the canopy (Figure 4.12). They begin to grow rapidly, however, when a canopy tree dies and leaves a hole admitting light into the understory. The juveniles begin to grow rapidly, but only the one that grows most quickly and strongly under the conditions of nutrient availability and soil type in the ecosystem will replace the original canopy tree. The others die out.

Water is a critical material for all plants. There is a strong selection for the more efficient in areas in which water is limiting. If individuals vary in their efficiency in absorbing water, the more efficient survive, and the less efficient do not. It is difficult to see this in a dense forest, or even in a semiarid or arid area where plants store quantities of water (as do the cacti, for example). However, woody vegetation in arid and semiarid areas may compete strongly for water. Their root systems may cover a larger area than their

Figure 4.12
Virgin sugar pine forest on Nevada Point Ridge, Eldorado National Forest, California. Note young sugar pine in the understory. [Photo courtesy U.S. Forest Service.]

Figure 4.13
Diagram showing the above-ground and below-ground extent of a desert plant.

above-ground growth (Figure 4.13), and water limitations minimize the overlap in roots of adjacent individuals. The result is a natural spacing of plants (Figure 4.14). The competition is powerful, but it takes place underground: As long as the roots can obtain water from a zone of the soil, they grow. When another plant's roots are there first, water limits further growth.

Figure 4.14
Vegetation on this range has been protected from cattle grazing for nearly 50 years. Still, little grass can grow because the area is fully occupied by mesquite. Note the regular pattern of vegetation in the foreground. [Photo courtesy U.S. Forest Service.]

As long as the root system can provide enough water (and nutrients), the plant survives. When it cannot, the plant dies, and the volume of soil formerly occupied by its root system is available to other plants.

TERRITORIALITY. Many patterns of social organizations operate in animal populations. They include those of bees, ants, and termites, which rival only urban people in their degree of division of labor and responsibility among individuals in the population. They also include other animals that show well-developed patterns during certain periods, such as the mating season, but much weaker patterns at other times. How do animals interact with each other, and why do they behave as they do? These questions are the basis of *sociobiology*.[29] This is an exceedingly complex and somewhat controversial science that can only be touched on briefly in a book of this type.

Many animals stake out a specific volume or area for their activities. These activities may include feeding, shelter, access to mates, sexual display, nesting, and so on. This *territory* is defended aggressively against invasion by other individuals of the same species and sometimes of other species. The key to territorial behavior is defense of the territory. This may be aggressive in some cases, involving actual bloodshed. More often, the confrontation is ritualized, with the threat of attack being sufficient to repel the invader.

The territory is different from, but related to, the *home range*. This is the area normally inhabited by an individual animal. In species that spend their lives defending strict territories, the home range and the territory may be all but identical. Other species share a home range with other conspecific individuals but maintain a distance from them. They defend a territory, but the territory may move with the animal.

Animals must recognize territorial boundaries, both their own and those of others. They must also be able to recognize each other. What this means depends on the kinds of animals involved. A bird can move quickly and has excellent vision. It does not need to mark territories explicitly, since it can fly to its boundaries when they are threatened, and it and its neighbors can recognize them by orienting themselves by other environmental features. A mammalian carnivore like a mountain lion, on the other hand, has a home range on the order of 13 to 65 km², and it cannot defend all of the boundaries at once as a discrete territory.[30] Mammals tend, therefore, to mark their territories with excrement, urine, or similar markers, and to recognize their conspecifics both by sight, and by these markings. Territories of these animals may overlap much more than those of birds, and sharing of home range by several individuals (but not at the same time!) is well known. The combination of a clear base, plus defined ways of marking position, allows animals to maintain a distance from one another. Most territorial animals move freely, but some attached species also show territoriality.[31]

Territoriality is a mechanism for dividing available space into more or less discrete subvolumes. Each territory reflects some resource that limits the population. As a result, the territory size of a given species may vary from place to place, as the resource base of the range varies (Figure 4.15). The dunlin, whose territory is sketched in this figure, is a sandpiper that lives in Alaska. Kolomak is at 61°N latitude. It is a subarctic site where food is relatively abundant and reliable. Population density of the dunlin is 30 pairs per hectare. Barrow is at 71°N latitude. It is arctic, with an unpredictable food supply and shorter summers. Dunlin density is about 6 pairs per

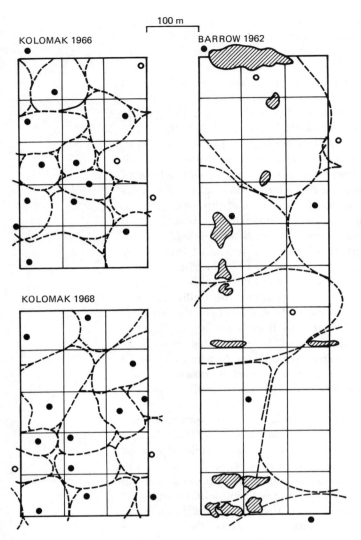

KOLOMAK 1966

100 m

BARROW 1962

KOLOMAK 1968

Figure 4.15
Territorial size of the dunlin, an Alaskan sandpiper at two localities. The population at Barrow, an arctic site, is ¹/₅ that at Kolomak, a subarctic site. Solid dots indicate nests; open dots indicate probable nest sites not uncovered. [Kolomak sites redrawn, with permission from Holmes, R. T., 1970. Differences in population density, territoriality, and food supply of dunlin on arctic and subarctic tundra in Watson, A. (ed.) *Animal Populations in Relation to their Food Resources* Oxford: Blackwell Sci. Pub.: 303–319. Copyright ⓒ Blackwell Scientific Publications. Barrow site adapted from Holmes, R. T., 1968. Breeding ecology and annual cycle adaptations of the red-backed sandpiper *(Calidris Alpina)* in northern Alaska. *Condor* 68, 3–46. Copyright ⓒ 1968, Cooper Ornithological Society.]

hectare.[32] Territories of great tits (a small bird, *Parus major*, related to the chickadee) were studied experimentally in Bean Wood, near Oxford (England). This wood maintained 14 pairs of tits in 1968. Half were shot (7 males, 7 females). Within 6 days, 13 birds from outside the Wood (5 pairs and 3 unpaired males) had replaced the birds removed. The next year, 6 pairs were removed from the 16 territories. Within 3 days, 4 new pairs had taken up residence (Figure 4.16). There does appear to be a correlation between resource availability in an area and the number of animals that can live there.[33]

Different species have different patterns of territoriality (Table 4.2). These intergrade into each other, but this classification points out some of the typical patterns found in nature and suggests the degree of differentiation. The first type corresponds to the normal image of territoriality (if, indeed, we have an image of it at all!). The territory includes all of the population's limiting resources, and makes an individual or breeding pair essentially self-sufficient within its territory. The second type does not include feeding

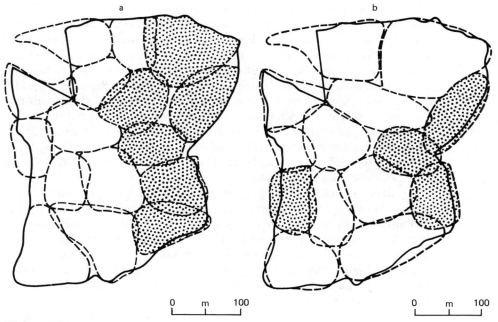

a   b

|0   m   100| |0   m   100|

Figure 4.16
Replacement of territorial birds by conspecific "floaters." Six pairs of great tit were re-
moved from the stippled territories in *a*. Within 3 days, 4 pairs had occupied the territo-
ries shown stippled in *b*. There had also been some readjustment of territorial boundaries.
[Redrawn, with permission from Krebs, J. R. 1971 Territory and breeding density in the
Great Tit (*Parus major L*). *Ecology* 52: 2–28. Copyright © Duke University Press.]

areas, but it does include reproductively important space. Mating and
breeding are carried out by such species within territories, but feeding may
be done in common feeding areas. The third type is characteristic of colonial
animals who feed over wide areas but breed in very restricted areas. Shore-
birds and dwellers on small islands are most typical of this type, but some

Table 4.2
Classification of Territories

Type	Defended Area	Resources Defended	Animals showing
A	Large	Shelter, Courtship Mating, Nesting, Food gathering	Benthic Fisher, Arboreal Lizards, Small Mammals, Insectivorous Birds
B	Large	Breeding activities	Some birds
C	Small	Nesting activities	Colonial birds, including seabirds, flamingos, ibis, herons, some finches Mud-dauber wasps, some bees
D	Small-Moderate	Communal mating areas	Insects and birds with communal courtship
E	Small	Roosting positions, shelters	Bats, socially roosting birds, spadefoot toads
F	Small	Space, nutrient supply	Colonial corals, active attached animals

A-E Based on E.O. Wilson, 1975. *Sociobiology: The New Synthesis*. Cambridge: Harvard University
Press.

insects that nest in aggregations but feed over wide areas also fit. The fourth type involves those animals that mate on well-defined *leks*, or mating grounds. This is a space where large numbers of animals come together and display to locate mates. The lek phenomenon is best known in large birds (e.g., grouse, sandpiper, peacocks, birds of paradise), but it is also found in fireflies, dragonflies, damselflies, and some ungulate mammals. The fifth type involves personal retreat. It is perhaps best developed for animals that live communally in restricted areas such as caves or trees, but it is found in other animals as well. It is very closely related to type C, except that the resource defended is personal space rather than nesting space. The final type is characteristic of several sessile, attached animals which actively maintain their positions on the substrate.

Not all of a population's range is equally desirable. Any individual who can derive a relatively greater portion of its resource needs from optimal areas obviously has a strong advantage over individuals who must gain their resources from more marginal areas. This is true for territorial and nonterritorial animals alike. Intraspecific competition in territorial animals may lead to the better territories being chosen by the more fit individuals. Once the territorial pattern is established, however, it is very difficult to break. An individual in the center of its own territory can generally defend it successfully against any conspecific invader, unless it is simply outclassed or ill.

The consequences of an animal's obtaining optimal territory should presumably be quite beneficial to it. Everything else being equal, one should expect the most fit individuals to occupy the optimal territories and to have the best reproductive success, thereby passing on relatively more genes to the next population than individuals in suboptimal territories. In red-winged blackbirds, the main determinant of male reproductive success is harem size, and female blackbirds tend to prefer experienced males. These tend, in fact, to occupy the better territories, but the female chooses the male, not the territory.[34] In bobolinks, polygynous males occupy territories that have a higher caterpillar density than those of monogamous males.[35] In both cases, then, polygyny was found in optimal habitats. Regardless of the specific mechanism of mates choosing each other or choosing their territory, the polygynous males defending the optimal territory had more offspring than monogamous males, and thereby contributed relatively more genes to the succeeding generation.

Territoriality is, in many ways, an expression of intraspecific competition, and its consequences are quite different at different densities. At low densities, there are enough optimal territories for everybody in the population. No individuals are excluded, and territorial considerations have little influence on reproductive success. At moderate densities, there is not enough optimal space for all, and some individuals are forced into marginal territories. Territoriality amplifies the genetic differences among individuals, so that the survival and/or reproductive success of those occupying optimal space is greater than that of those occupying marginal space. In either case, the genetic contribution of those individuals to subsequent generations is maximized. At high densities, there are not even enough marginal territories to go around, and some individuals are forced to become "floaters," without any territory at all. Their reproductive success is much lower (if they are successful at all) than that of individuals with territories, and their mortality is higher. Nevertheless, they buffer the population in that they can move into a territory that has been vacated by its tenant (e.g., by death, as with the great tits in the experiment cited previously).

Why should territoriality have been selected for in the course of evolution? In populations that are limited by their resource base (e.g., by food supply or specialized nesting sites, as opposed to regulation by predators or parasites) territoriality is one way for a population to regulate itself. Because the mortality and reproductive failure of individuals excluded from territories tend to be substantially greater than those of individuals who have territories, genes conferring territoriality are strongly favored by natural selection as long as the energetic cost of defending territories is not unreasonably high. In fact this is typically the case. If anything, the tendency of confrontation to be ritualized rather than violent makes territorial defense less energy-consuming than overt, aggressive competition. In many species, holders of adjacent territories may compete aggressively to establish their boundaries but waste very little energy on confrontation thereafter. They recognize each others' calls or other markers but do not respond to normal movement as to an invasion. A non-adjacent conspecific, however, would elicit an aggressive response if it approached the territorial boundary. Territoriality is thus a powerful mechanism for buffering a population and relating it to resource requirements.

SOCIAL DOMINANCE. Territoriality is a mechanism that establishes a spatial order among individuals in a population. Social dominance is analogous to this in many ways, except that its organization is more abstract. In some species, one individual is a despot who lords over the others in the population, and the other members are all of equal rank. More commonly, populations show a network of rank orders, in which each individual has a definite rank, and a large number of ranks exists.

The mechanism for social dominance as an evolutionary phenomenon is very similar to that for territoriality. High-rank individuals are more likely to get adequate food, mate effectively and often, and pass their genes along to the next generation than are low-rank individuals. Its most familiar (and perhaps dramatic) manifestation is the *pecking order* (the name is derived from the social order by which domestic barnyard fowl—which have rather well-developed pecking orders—peck at each other without fear of reprisal). Each individual in the population is known to all others. The order in which they feed or obtain nesting sites depends on their rank, with individuals of low rank deferring to those of higher rank. When food and nesting sites are in adequate supply for the entire population, all individuals have access to them. When they are not, only the animals at the top of the hierarchy are successful. The rest starve, are excluded from nesting sites, or are inhibited in their sexual development.

As examples, dominant wood pigeons (*Columba palumbus*) feed more quickly than subordinates, and they are found at the center of a feeding flock. Because of their location in the center, they do not have to look back at the rest of the advancing flock or out at potential predators. In fact, subordinate birds may be able to eat enough to last only a single night, and they are in danger of death if the temperature drops sharply, or bad weather prevents feeding the next morning.[36] An experiment using laboratory mice demonstrated that the dominant males (comprising one third of the population) sired 92% of the offspring. In 82% of the instances, they sired all of the offspring.[37]

Social dominance is found in many different groups of animals. It controls several resources, and it is correlated with many factors. It is found in insects (some beetles, wasps, bumblebees, stingless bees, ants), crustaceans (spiny lobsters and crayfish), fish (sunfish, swordtails, and minnows), reptiles (Galapagos tortoise), birds (grouse and boobies), rodents (voles and deer mice), ungulates (alpacas, llamas, guanacos, mountain sheep, dairy cattle, axis deer, some antelopes), and primates (macaques and mangabeys). Resources controlled by the dominance orders include females, food, oviposition rights, nest sites, space, and resting places. It is correlated with size, arrival sequence, state of ovary development, victory or defeat in previous encounters, age, experience, and timing of the reproductive cycle. Correlation can be very strong or fairly weak.[38]

Considering how similar the implications of territoriality and social dominance are, it is not surprising that both occur in some species. When green sunfish (*Lepomis cyanellus*) are raised in small aquaria, they organize themselves into a clear pecking order, but the males will stake out territories if they are less crowded.[39] As might be expected, the dominant males get the choicer territories, and the lowest-ranking males may or may not get territories. Some crayfish are normally territorial, but they too form dominance hierarchies at high densities.[40]

INFORMATION AND INTRASPECIFIC BEHAVIOR. The critical features of intraspecific behavior for population organization are that they are related to the resource requirements of the species, that they require individual recognition of individuals by other members of the population, and that they result in powerful genetic advantage for the dominant individuals or those occupying optimal territories. They also have a role in population regulation, but that is incidental.

Social interactions are found in such a broad cross-section of the animal kingdom that they are likely to be typical of higher animals in general. The fact that social organization requires that individuals be able to recognize one another (by sight, call, smell, marking, or other means) has a corollary of great significance to the study of ecosystems. Social behavior is based largely on an information network that interlinks individuals. Animals make the responses they do because they receive low-energy signals that are interpreted by the brain. The same is true of many of the interactions involved in interspecific interactions as well, but it is not as universal (and does not affect plants at all).

As a result, the organization of ecosystems can revolve around two distinct (but related) networks. The first is the primary network of energy and materials flowing through the ecosystem and providing the resource needs of the various life forms. The second is information that provokes much of the behavior of animals in the ecosystem. The first is cumbersome and slow; the second is much more efficient and rapid. Population regulation in real ecosystems depends on both of these networks, and it is not at all obvious how they work together. A given population is influenced by other species and by other members of its own. It is part of positive and negative feedback loops, some of which are concrete-material based and some of which are information based. Most of our sophisticated analytical tools now in use are not intended to deal with this level of complexity. We really do not know how much difference it makes that ecosystems are organized as they are. But we need to understand and accept the complexity as a fundamental given in the system.

It is sometimes difficult, for example, to identify what forces are at work motivating a given phenomenon. In the example of amensalism discussed earlier, there is also evidence that the exclusion of grass from the vicinity of *Salvia leucophylla* and *Artemisia californica* is not due to volatile materials produced by the vegetation but rather to herbivory by rodents living in the shrubs. The evidence is not totally clear in either direction, and it is quite likely that both effects are involved, at least to some degree.[41]

Another notion that is becoming increasingly clear is that it is sometimes very difficult even to discuss population phenomena without looking in some detail at other members of the community that are closely associated with the species of primary interest. This does not mean doing a complete analysis of the community; it means simply noting that certain groups of species (sometimes termed *guilds*) are extremely closely related in how they all respond to environmental stimuli, and that the responses of any one member of the guild would be very different in the absence of the other members.[42]

Notes

[1] Andrewartha, H. G., and Birch, L. C., 1954. *The Distribution and Abundance of Animals*. Chicago: University of Chicago Press.

[2] Rosenzweig, M. L., 1981. A theory of habitat selection. *Ecology*, **62**, 327–335.

[3] Hardin, G., 1960. The competitive exclusion principle. *Science*, **131**, 1292–1297; Jaeger, R. G., 1974. Competitive exclusion: Comments on survival and extinction of species. *BioScience*, **24**, 33–39.

[4] Gause, G. F., 1934. *The Struggle for Existence*. Baltimore: Williams and Wilkins; Bodenheimer, F. S., 1938. *Problems of Animal Ecology*. London: Oxford University Press.

[5] Tilman, D., 1977. Resource competition between planktonic algae: an experimental and theoretical approach. *Ecology*, **58**, 338–348.

[6] Tilman, D., 1980. Resources: a graphical-mechanistic approach to competition and predation. *Am. Nat.*, **116**, 362–393.

[7] Park, T., 1962. Beetles, competition, and populations. *Science*, **138**, 1369–1375.

[8] Bovbjerg, R. V., 1970. Ecological isolation and competitive exclusion in two crayfish (*Orconectes virilis* and *Orconectes immunis*). *Ecology*, **51**, 225–236.

[9] Williams, A. H., 1980. The threespot damselfish: A noncarnivorous keystone species. *Am. Nat.*, **116**, 138–142; Williams, A. H., 1981. An analysis of competitive interactions in a patchy back-reef environment. *Ecology*, **62**, 1107–1120.

[10] Ehrenfeld, D. W., 1970. *Biological Conservation*. New York: Holt, Rinehart and Winston, Inc.

[11] Schemske, D. W., 1981. Floral convergence and pollinator sharing in two bee-pollinated tropical herbs. *Ecology*, **62**, 946–954.

[12] Beattie, A. J., and Culver, D. C., 1981. The guild of myrmecochores in the herbaceous flora of West Virginia forests. *Ecology*, **62**, 107–115.

[13] Heithaus, E. R., 1981. Seed predation by rodents on three ant-dispersed plants. *Ecology*, **62**, 136–145.

[14] Stiles, E. W., 1980. Patterns of fruit presentation and seed dispersal in bird-disseminated woody plants in the eastern deciduous forest. *Am. Nat.*, **116**, 670–688.

[15] Temple, S. A., 1977. Plant-animal mutualism: Coevolution with Dodo leads to near extinction of plant. *Science*, **197**, 885–886. See also comment and response in *Science*, **203**, 1363–1364.

[16] Janos, D. P., 1980. Vesicular-arbuscular mycorrhizae affect lowland tropical rain forest plant growth. *Ecology*, **61**, 151–162.

[17] Stewart, W. D. P., 1977. Present-day nitrogen-fixing plants. *Ambio*, **6**, 166–173.

[18] Whittaker, R. H., 1970. The biochemical ecology of higher plants. Chapter 3 in Sondheimer, E., and Simeone, J. B., *Chemical Ecology*. New York: Academic Press, Inc., 43–70; Whittaker, R. H., and Feeny, P. P., 1971. Allelochemics: Chemical interactions between species. *Science*, **171**, 757–770.

[19] Muller, C. H., 1966. The role of chemical inhibition (Allelopathy) in vegetational composition. *Bull. Torrey Bot. Club*, **93**, 332–351.

[20] Rice, E. L., and Pancholy, S. K., 1972. Inhibition of nitrification by climax ecosystems. *Am. Jour. Bot.*, **59**, 1033–1040; Rice, E. L., 1974. *Allelopathy*. New York: Academic Press.

[21] DeBach, P., 1974. *Biological Control by Natural Enemies*. New York: Cambridge University Press.

[22] Doutt, R. L., 1964. The historical development of biological control. Chapter 2 in DeBach, P., *Biological Control of Insect Pests and Weeds*. New York: Reinhold Publishing Corporation, 21–42.

[23] Hairston, N. G., Smith, F. E., and Slobodkin, L. B., 1960. Community structure, population control, and competition. *Am. Nat.*, **94**, 421–425.

[24] Holloway, J. K., 1964. Projects in biological control of weeds. Chapter 23 in DeBach, P., *Biological Control of Insect Pests and Weeds*. New York: Reinhold Publishing Corporation, pp. 650–670.

[25] Mech, L. D., 1966. The Wolves of Isle Royale. *U.S. Nat. Park Fauna*, **7**; Jordan, P. A., Botkin, D. B., and Wolfe, M. L., 1971. Biomass dynamics in a wolf population. *Ecology*, **52**, 147–152.

[26] Owen, D. F., 1980. How plants may benefit from the animals that eat them. *Oikos*, **35**, 230–235.

[27] Whittaker, R. H., and Feeny, P. P., 1971. Allelochemics: Chemical interactions between species. *Science*, **171**, 757–770.

[28] Gilbert, L. E., 1971. Butterfly-plant coevolution: Has *Passiflora adenopoda* won the selectional race with heliconiine butterflies? *Science*, **172**, 585–586; Rathke, B. J., and Poole, R. W., 1975. Coevolutionary race continues: Butterfly larval adaptation to plant trichomes. *Science*, **187**, 175–176; Pillemer, E. A., and Tingey, W. M., 1976. Hooked trichomes: A physical plant barrier to a major agricultural pest. *Science*, **193**, 482–484.

[29] Wilson, E. O., 1975. *Sociobiology: The New Synthesis*. Cambridge, Mass.: Harvard University Press.

[30] Hornocker, M. G., 1969. Winter territoriality in mountain lions. *Jour. Wildlife Mgmt.*, **33**, 457–464.

[31] Richardson, C. A., Dustan, P., and Lang, J. C., 1979. Maintenance of living space by sweeper tentacles of *Montastrea cavernosa*, a Caribbean reef coral. *Marine Biology*, **55**, 181–186.

[32] Holmes, R. T., 1970. Differences in population density, territoriality, and food supply of dunlin on arctic and subarctic tundra, in Watson, A., ed. *Animal Populations in Relation to Their Food Resources*. Oxford: Blackwell Scientific Publications, 303–319.

[33] Krebs, J. R., 1971. Territory and breeding density in the great tit, *Parus major* L. *Ecology*, **52**, 2–22.

[34] Yasukawa, K., 1981. Male quality and female choice of mate in the red-winged blackbird (*Agelaius phoeniceus*). *Ecology*, **62**, 922–929.

[35] Wittenberger, J. F., 1980. Vegetation structure, food supply, and polygyny in bobolinks (*Dolichonyx oryziforus*). *Ecology*, **61**, 140–150.

[36] Murton, R. K., Isaacson, A. J., and Westwood, N. J., 1966. The relationships between wood pigeons and their clover food supply and the mechanism of population control. *Jour. Appl. Ecol.*, **3**, 55–96.

[37] DeFries, J. C., and McClearn, G. E., 1970. Social dominance and Darwinian fitness in the laboratory mouse. *Am. Nat.*, **104**, 408–411.

[38] Wilson, E. O., 1975. *Sociobiology: The New Synthesis*. Cambridge, Mass.: Harvard University Press.

[39] Greenberg, B., 1947. Some relations between territory, social hierarchy, and leadership in the green sunfish (*Lepomis cyanellus*). *Physiol. Zool.*, **20**, 267–299.

[40] Lowe, M. E., 1956. Dominance-subordinance relationships in the crawfish *Cambarellus shufeldti*. *Tulane Studies in Zoology, New Orleans*, **4**, 139–170; Bovbjerg, R. V., 1956. Some factors affecting aggressive behavior in crayfish. *Physiol. Zool.*, **29**, 127–136.

[41] Bartholomew, B., 1970. Bare zone between California shrub and grassland communities: The role of animals. *Science*, **170**, 1210–1212. See also comment and response in *Science*, **173**, 462–463.

[42] Atsatt, P. R., and O'Dowd, D. J., 1976. Plant defense guilds. *Science*, **193**, 24–29.

The Growth and Regulation of Populations

5

THE RULES GOVERNING population growth are far from simple. Many interactions are involved, of which some are internal to the population and governed by its genetic basis, some depend on interspecific interactions, and some are based on factors of the abiotic environment. Depending on conditions, population density may be constant, or it may fluctuate widely. If a new species is introduced into a new area, its initial density may be far below the level that the environment is capable of supporting, and population growth may be rapid. Conversely, if the environmental factors change, the density of the population may drop.

Population Dynamics

Population dynamics is the study of the growth and regulation of populations. It represents a tension between the tendency of a population to grow and the limits to that growth imposed by the environment. For instance, if a single oyster can produce 1 million (10^6) eggs per year (this figure is not unreasonable and may even be low), if no limits to the growth of the oyster population were set by the environment, and if all the offspring lived and reproduced once, then the oyster population would grow as shown in Table 5.1. To put this into perspective, if an average oyster weighed 500 g, the number of oysters produced after five generations would weigh five hundred times the weight of the sun. Obviously, the constraints imposed by the environment prevent this from happening, but the tendency of populations to grow at an exponential rate given the lack of environmental restraints is a characteristic of all populations, even those whose production of young is nowhere near as prolific as that of oysters. This tendency is sometimes called the population's *biotic potential*. When populations do not show unlimited growth, it is because of the *environmental resistance* to growth.

Vital Rates

The growth of populations can be described in terms of the *birth rate*, the *death rate*, and the number of individuals in the population, as well as migration of individuals into or out of the area. The birth rate and death rate are often collectively termed the *vital rates* of the population, and the *net growth rate* is equal to the birth rate minus the death rate. The number of individuals in the population may be considered either in terms of population size or density.

108

Table 5.1
Growth of Population of Oysters Assuming
Each Female Produces 1 Million (10^6)
Daughters, and Each Lives to Reproduce,
in the Lack of Any Environmentally Induced
Limitations on the Population

Generation	Number of Female Offspring
0	1
1	10^6
2	10^{12}
3	10^{18}
4	10^{24}
5	10^{30}

The most straightforward way of measuring the vital rates of populations is the so-called *crude rates* of birth, death, and growth. These do not take into account the effects of age distribution, sex ratio, or migration, but they are adequate for many purposes. The reader is referred to texts in demography for a discussion of more sophisticated measures of vital rates.[1] The crude rates are defined as follows:

$$\text{Birth rate } (b) = \frac{\text{number of births per unit time}}{\text{average population}}$$

$$\text{Death rate } (d) = \frac{\text{number of deaths per unit time}}{\text{average population}}$$

$$\text{Growth rate } (r) = \frac{\text{number of births} - \text{number of deaths}}{\text{average population in time interval}}$$

The actual change in the population numbers (ΔN) over any span of time (Δt) is equal to rN. This can be written as $\Delta N/\Delta t = rN$. It is somewhat easier to use notation from calculus and to think in terms of the rate of change at any instant of time. The shorthand for this is (dN/dt), which differs conceptually from $\Delta N/\Delta t$ only in that the time interval considered is infinitely short. In any case, the instantaneous rate of change of the population can be expressed as $dN/dt = rN$. Let us now say that the population includes N_t individuals at some arbitrary time t, and that it contained N_0 individuals at the beginning (time t_0). Calculus will allow us to state that the size of the population at time t is expressed by the equation $N_t = N_0 e^{rt}$, where $e = 2.71828 \ldots$, the base of the natural logarithms.[2]

If r were constant, population growth would be exponential, as in the oysters mentioned earlier. But no population can grow exponentially very long, and r is never constant. Because r is the difference between the birth rate and the death rate, changes can result from variation in either or both of those rates. For many organisms, the birth and death rates are related to the population density. If density is high, the birth rate is low, because of inadequate nutrition or other aberrations associated with crowding. The birth rate tends to be higher at lower densities, except that individuals may have trouble finding each other and mating if the density is extremely low. There is often an optimum density at which the birth rate is maximized. This is often relatively low, but it need not be.

The relationship between mortality and density is similar. Mortality is almost always highest at very high densities because of the hazards of overcrowding and the likelihood of concentrated predation or spread of disease, but it may be quite high at very low densities as well, since several individuals of a given species are often better able to survive a period of stress than is a single individual. For instance, the heat generated by a cluster of bees is sufficient to allow survival of the cluster in temperatures low enough to kill the bees, were they not clumped together. There is no general curve to relate vital rates to population density for all species. Many of the factors that influence a population's vital rates show a relationship to density, but many do not. These matters are best regarded as a property of any given population, considering the characteristics of the ecosystem in which it is found.

To visualize the relationship between population growth, birth rate, and death rate, let us consider two very simple examples. In both cases, the birth rate is constant ($b = b$) for all population densities. In the first case, the death rate is also constant ($d = d$). It may be either higher or lower than the birth rate (Figure 5.1). The first case represents what is sometimes called *density-independent* population regulation. Whether the population is expanding ($b > d$, so that r is positive) or declining ($b < d$ so that r is negative) has no relation to the number of individuals in the ecosystem. The population shows either an exponential increase to infinite density or an exponential decline to extinction.

Of course, no population can grow exponentially forever. But conditions in many ecosystems allow b to be substantially larger than d for a period of time, following which they change so that d becomes much larger than b. The response of populations to variations of this sort is an exponential population explosion during favorable conditions, followed by a "crash" when conditions change. Examples include insect outbreaks in crops or "blooms" of algae in a polluted lake. Figure 5.2 shows the boom and bust cycles of diatom populations in Lake Michigan at different times of year, triggered by variations in the abiotic factors within the lake.

The second case of population growth also has a constant birth rate, but the death rate increases as a linear function of population density ($d = aN + c$). It represents a population in which mortality is *density-dependent*. This indicates a negative feedback mechanism that controls mortality through factors such as the availability of food (e.g., Figure 1.5), or through a more

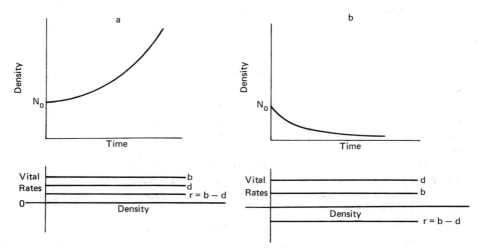

Figure 5.1
Population growth under two instances of constant birth and death rates. a. Birth rate > death rate; therefore growth rate is positive. b. Birth rate < death rate; therefore growth rate is negative.

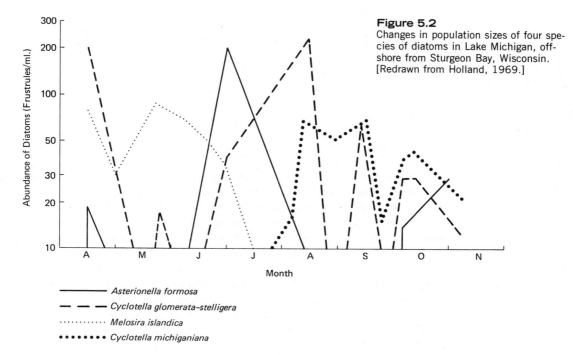

Figure 5.2
Changes in population sizes of four species of diatoms in Lake Michigan, offshore from Sturgeon Bay, Wisconsin. [Redrawn from Holland, 1969.]

——— Asterionella formosa
— — — Cyclotella glomerata-stelligera
··········· Melosira islandica
•••••••• Cyclotella michiganiana

complex factor or set of factors related to population density. As shown in Figure 5.3, the population rises or falls from its initial density, N_0, to the equilibrium level in a fashion described by an S-shaped, or *logistic curve*.

Logistic growth is characterized by an equilibrium density, or *carrying capacity*. This is represented in the figure by the letter K. To distinguish between actual growth rate and potential growth rate, it is often useful to define an equation in which biotic potential and environmental resistance are shown explicitly. Biotic potential, or *intrinsic rate of natural increase*, can be symbolized r_0 and represents the growth rate of a population that is infinitely small. It is thus the rate at which the population would grow, were there no environmental resistance in the ecosystem. Environmental resistance is related to the carrying capacity. In this particular case, is is equal to $(K - N) / K$. Thus, the overall growth of the population can be described by the equation, $dN/dt = r_0 N [(K - N) / K]$. This is commonly known as the *logistic equation*, and it is the most widely used simple model to describe regulated population growth.

The assumptions that birth rate is independent of population density and that the death rate increases steadily with density are probably never true in natural ecosystems. Any other set of assumptions that give a linear relation-

Figure 5.3
Population growth when birth rate is constant, but death rate is a linear function of population density.

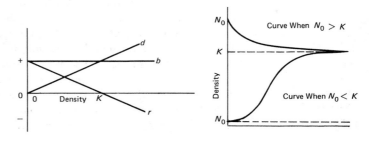

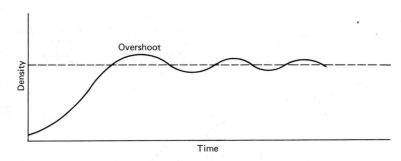

Figure 5.4
Typical growth of populations to equilibrium, showing oscillation around equilibrium density.

ship between net growth rate and population density are equally oversimplified. Nevertheless, the simple logistic equation is sometimes sufficient to describe the growth of populations to equilibrium, and it has formed the basis of a great deal of theoretical work in population ecology. For instance, logistic growth characterized the experimental populations of *Paramecium* studied by Gause (Figure 4.2). But this is the exception, not the rule. Few populations in the field are static: Birth rate and death rate both fluctuate in response to the constellation of environmental factors to which real populations respond. The equilibrium level *K* may itself be a function of the season of the year or some other environmental factor, and vary accordingly. Finally, the feedback mechanisms that regulate populations are never perfect. Populations that appear to show logistic growth almost always show some degree of fluctuation about the equilibrium density (Figure 5.4).

Life History Strategies

Different populations are characterized by different *life histories*. The evolutionary significance of life history is not fully understood, and it has become a matter of some controversy in theoretical ecology, but some observations regarding the life history phenomenon are commonly accepted.[3]

Some environments are temporary or unpredictable, and long-term equilibria may not be possible. Species inhabiting these environments must be *opportunists* that can locate a suitable habitat, rise rapidly to a high density, and then either locate another suitable habitat after optimal conditions have passed or survive periods of stress at very low densities or in a dormant state. It makes little sense to speak of equilibrium population size, because the survival of organisms in such ecosystems is most closely related to their ability to seize opportunities from the random or unpredictable variation in environmental factors. Opportunism in species is commonly characterized by a potential for a high rate of population growth. Individuals tend to mature early, expend a great deal of energy on reproduction, and produce a large number of offspring. Organisms in the first group can colonize an area quickly and grow to high densities, and they can maintain a population in a fluctuating environment when adult mortality is high.

Other species tend to maintain stable equilibrium populations over an extended period of time. They are adapted to compete successfully at more or less constant densities, regardless of minor variations in environmental factors. They are "settled" species, regulated by well-developed feedback mechanisms with the equilibrium population density at or near the carrying capacity. They mature later than opportunists, expend relatively less energy on reproduction, and produce fewer offspring. They tend to be found in more stable environments or fluctuating environments in which adult mortality is relatively low.

Species show a gradation from the most opportunistic to the most settled, and it is not very useful to try to force all species into these pigeon holes. It is more important for our purposes that the gradient exists than to classify all organisms. The difference between opportunistic and settled species has been demonstrated many times on empirical grounds, however, and it is a useful distinction.

From an ecosystem perspective, the main difference between an opportunist characterized by rapid exponential growth and a settled species characterized by logistic growth is the existence of an negative feedback mechanism capable of regulating growth and based on the density of the population. The precise details of the feedback loop may be different in every instance. How a population responds to stimuli depends, among other things, on its genetic base. The feedback mechanism regulating populations includes not only these responses, however, but also the environmental factors generating the stimuli and any secondary stimuli generated by the population's responses. We do not totally understand the nature of the adaptations that lead one species to be an opportunist and another to be settled. To do so, we would have to be able to design coupled experiments and field observations that are presently beyond our capabilities.[4]

The evidence for these feedback devices is compelling, however. Some populations show minimal fluctuation about an equilibrium density, indicating an effective feedback control mechanism of some sort. Others oscillate around an equilibrium, indicating that feedback control operates in the population, but without an effective mechanism for damping out variation induced either by environmental fluctuation or by the population itself. Finally, some species show no evidence of an equilibrium population density, suggesting that they may not be feedback-regulated (Figure 5.5). Even a single species may show some variation from place to place in the effectiveness of their regulation. For example, feedback may be effective in the cen-

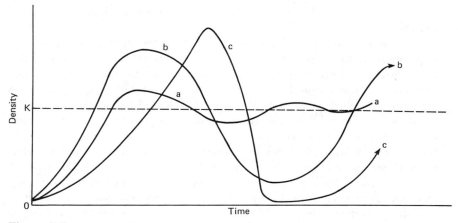

Figure 5.5
Three curves of population growth: a. Equilibrium level exists, and control around that equilibrium is fairly tight. b. Equilibrium level exists, and feed back control exists to correct for "overshoot", but there is no tight control around equilibrium. c. No evidence for equilibrium population density.

ter of the range but less so at the edges. As a result, the population may be stable in one place but oscillate or even become locally extinct during times of stress in the marginal areas, forcing it to recolonize the area after the stress period has passed.

A species must accommodate itself to many different facets of its environment, and the strategy of its life history is a significant type of adaptation. Environmental fluctuation, characteristics of the resource base, types of predators, and nature of potential food sources all represent factors that differ from place to place and from time to time. Different strategies of reproductive energy expenditure, growth, degree of feedback control, etc. are appropriate for different situations, and are subject to natural selection.

Density Dependence and the Regulation of Population Size

Feedback mechanisms involved in regulating population density operate relative to the density itself: The operative environmental factor impinges on the population differently at different densities. Many interspecific and intraspecific factors operate in this way, although few abiotic and some biotic factors do not. The terminology that has grown over the years differentiates *density-dependent* (i.e. feedback-based) mechanisms from *density-independent* (i.e. feedback-unrelated) mechanisms. Natural populations are subject to both forms of regulation to a greater or lesser extent, although the multitude of factors comprising the environment makes it difficult to predict the precise relationship between the complex of environmental factors and the density of the population.

In many instances, notably those in which random variations in the environment trigger a rise in population size (as in the case of desert plants whose seeds may lie dormant for long periods of time but sprout rapidly after a heavy rain), the density of the population is essentially irrelevant, and a given percentage of the organisms will grow and reproduce, regardless of their total number. Likewise, when a random variation in the ecosystem causes a rapid decline in population size, as with a flood or forest fire, the percentage losses within a population are independent of the population's density.

There is a connection between life history and the significance of density-dependent control in populations, but it is not a firm one. Feedback control is important in settled species. They could not have constant populations without regulation around an equilibrium level, and their adaptations to low reproductive effort would not otherwise be selected for. Opportunists may not be subject to feedback control, or only intermittantly. Density-dependent factors may operate, but they are so imperfect that the population does not reach an equilibrium level, but rather fluctuates widely around a median density, so that a high intrinsic rate of increase is necessary for survival.

An example of feedback control in an opportunistic species is shown in an experiment involving the herbivorous mite, *Eotetranychus sexmaculatus*, and the predatory mite, *Typhlodromus occidentalis*.[5] *E. sexmaculatus* was introduced into the experimental chamber, with several oranges as its food supply. Once it was established, *T. occidentalis* was also introduced and allowed to feed on *E. sexmaculatus* (Figure 4.10). The design of the particular experiment shown in this figure provided enough heterogeneity in the landscape that *E. sexmaculatus* could hide. Initially, *E. sexmaculatus* increased rapidly in density. This represented an increase in the food supply for *T. occidentalis*, which responded by increasing its own population. An increase in the predation pressure drove down the population of *E. sexmaculatus*, and the corresponding decrease in its food supply drove down the population of *T. occidentalis*, thus lowering predation pressure and allowing *E. sexmaculatus* to expand again.

There were several versions of this experiment. Most did not have enough refuges to allow *E. sexmaculatus* to regroup its population and begin to rise again. As a result, the most common result was an explosive increase in the herbivore species, occasioned by introduction into an "empty" food-filled environment. Once the *E. sexmaculatus* population had built up to a sufficient level, *T. occidentalis* was able to explode for precisely the same reason. However, *T. occidentalis* was a sufficiently efficient predator to drive *E. sexmaculatus* into local extinction in most versions, with its own population "crashing" soon thereafter. When refuges for *E. sexmaculatus* and provision for migrating from one orange to another were provided, however, the *T. occidentalis* could drive *E. sexmaculatus* into extinction on one orange, but the herbivore would be initiating its explosive rise on another. The result was an oscillation that was very unstable locally, but fairly stable over the whole experimental ecosystem.

Even the simplest of natural ecosystems is much more complex than a laboratory experimental chamber, so that refuges for species under pressure are much more likely. If the experimental chamber had been an entire orange grove (instead of a chamber with 27 oranges in it), it is entirely possible that the oscillation could have persisted indefinitely.

Grain

This example points out more than the way in which the responses of two mites lead to an oscillation in their mutual frequencies. It also points out the critical role of environmental heterogeneity in determining not only the outcome of the interspecific factors discussed in the previous chapter, but also the course of feedback control of ecosystems as a whole. This notion has been described graphically as the *grain* of an ecosystem.

Consider, if you will, a landscape as though it were a piece of sandpaper. At the extremes, it can be *fine-grained*, with a smooth surface on which any one place is like any other, or it can have a *coarse-grained*, surface containing many different kinds of microenvironments (Figure 5.6). There are few (if any) refuges in a fine-grained environment, but there is a broad diversity in a coarse-grained environment.

Perhaps the quintessential fine-grained ecosystem is the open deep sea. It is evenly dark, virtually endless, featureless, and uniform. The temperature is low and constant. The characteristics of any point in it are identical to any other at the same depth. The only dimensions that can be perceived by an organism living in the deep sea are vertical, based on pressure, and the direction of current flow (if any).

Contrast this with a wooded public park. It contains several species of trees, each of which provides different habitat for different species. The sunny side of each tree is warmer and dryer than the shady side, and there

Fine Grain

Coarse Grain

Figure 5.6
Difference between a "fine-grained" environment and a "coarse-grained" environment.

are crevices in the bark for insects and other small animals to hide in. The upper side of each leaf is a different kind of place from the lower side, and the leaves that are open to the sun have different characteristics from leaves that have grown in the shade, even on the same tree. Different plants grow in the shrub zone than grow in the canopy or in the herbaceous layer next to the ground, and the soil is almost another world when compared with the above-ground environment. One could easily list hundreds of different kinds of places for organisms to live in even a simple suburban wooded park. The number would be much higher in a virgin temperate forest or a tropical rainforest. It would also be higher in a coral reef or a swamp.

The significance of grain is that it has a tremendous influence on the behavior of ecosystems. An oscillation that would be unstable in a fine-grained environment may be stable in a coarse-grained ecosystem. Likewise, what is fine-grained for one species (such as a bison) may be coarse-grained for another (such as a meadow mouse). In general, coarse grain confers stability on an environment.

Evolution of Population Regulation

The factors that regulate populations are no more constant than the populations themselves. They vary with climate and other fundamental abiotic signals. The biological factors also change as the result of evolution. Natural selection tends to alter the genetic variability of a population as certain individuals pass their genes on to the next generation more than do others. In general, genes that confer an advantage to the individuals bearing them increase in frequency, while those that confer a disadvantage tend to decrease. Because of the great heterogeneity of most ecosystems, different genes may confer advantage or disadvantage at different times and places, so the variability of a typical natural population remains high.

Two closely associated populations tend to exert a measure of control over each other not only with respect to population regulation, but also evolutionarily. Whether they constitute a mutualistic association, a predator-prey pair, or competitors, a change in gene frequency in one is likely to result in evolution of the other as well. The results can be quite striking. One particularly interesting manifestation of this is what David Pimentel calls *genetic feedback*.[6] A classic example discussed by Pimentel involves the European rabbit (*Oryctolagus cuniculus*), which was introduced into Australia at the end of 1859. The subsequent population explosion that turned this species into a major pest is well known.

A strain of *Myxoma* virus was obtained from South American rabbit populations and introduced into Australia in 1950. The resulting myxomatosis epidemic proved fatal to 97–99% of the rabbits. A second outbreak of myxomatosis was fatal to 85–95% of the population, and third outbreak was fatal to 40–60%. The decreasing effectiveness of the virus suggests that evolution was occurring in both populations, with the rabbits becoming less susceptible to the disease and the virus strain becoming less virulent. The selective advantage to the rabbits of lowered susceptibility to disease is obvious: Tolerance of a fatal disease does indeed lead to greater survival. The mechanism of the genetic shift in the rabbit population is also apparent. Some individuals were more resistant to the disease than others, due to normal genetic variation. These survived and passed their myxomatosis-resistant genes to succeeding generations.

It is not as obvious how loss of virulence was selectively advantageous to the virus, but this appears to have been the case. It is clearly prejudicial to the continuity of a consumer population if it is so successful that it destroys its food supply, since it destroys its means of survival in doing so. A reduction in the virulence of a disease ensures a greater likelihood of long-term continuity of the food supply and therefore of the virus. But one must document a mechanism as well as an advantage.

Two mechanisms to select for lowered virulence were available in the case of myxomatosis and rabbits. First, the *Myxoma* virus is spread in Australia by *Aedes* and *Anopheles* mosquitoes, which feed only on living animals. Because a rabbit infected by a less virulent strain lives longer than one infected by a highly virulent strain, the probability of the mosquitoes' feeding on sick (but alive) rabbits infected with less virulent *Myxoma* virus is substantially greater than that of their encountering more deadly strains in living rabbits. Secondly, if there is any geographic variation in myxomatosis virulence, the areas characterized by lower virulence can support more rabbits, and hence a higher density of virus as well. Eventually, a balance is struck by the maximum effectiveness of the virus strain's ability to convert rabbit protoplasm into virus and the continuity of the viruses' food supply.

A similar pattern of coevolution can be expected in other interacting populations. As shown in the discussion of the yucca moth (page 89), coevolution can affect mutualism as well as predation. It can even take place in a highly integrated community whose populations are regulated through interactions of many species. In this case, of course, natural selection would operate throughout the entire species complex.

A caveat should be entered here regarding natural selection in real populations. Most textbooks (including this one) talk as though the forces of natural selection were relatively simple, and that they acted all the time, or at least most of the time. This is not the case. Selection can act very much more strongly at some times than at others. *Bottlenecks* may provide the most significant stimuli for natural selection, and these bottlenecks may be very infrequent.[7] Figure 5.7, for example, documents the soil moisture deficit at the Royal Botanical Gardens at Kew, in southeast England. This is a measure of stress that can be very significant to plants. The critical level of 84mm. shown in the figure is somewhat arbitrary, but it is not an unreasonable one. A similar picture would emerge if it were 80mm. or 90mm. The point is that

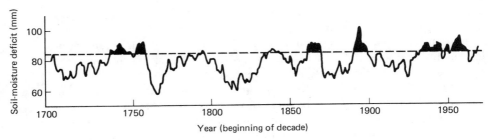

Figure 5.7
The soil-moisture deficit at Kew, England, has been averaged over the growing seasons of a 279-year period. The plotted values are 10-year running means, and thus the value for any given year is the average deficit for that year and the 9 following years. In this analysis, then, severe soil-moisture stress is considered in terms of averages over a decade rather than single extreme years. An arbitrary level of 84 mm is shaded to accentuate periods of higher deficits. [Reprinted by permission from *Nature,* Vol. 265, No. 5593, pp. 431–434. Copyright © 1977, Macmillan Journals Ltd.]

the periods of elevated and extended water stress that would make a real difference in plant survival (and hence affect the passage of genes influencing stress resistance from one generation to the next) occur at intervals often separated by decades. It would be entirely possible to study an ecosystem in the field for many years without seeing the bottlenecks that have the greatest influence on the populations.

Age

With respect to population dynamics, all individuals in a population are not equal. Some cannot reproduce, because they are too young or too old. Organisms are much more likely to die at some ages than at others, just as they are more likely to reproduce at certain ages. The *distribution* of ages in a population may remain fairly constant, or it may change throughout the generation time. A population's age distribution results from the overall set of factors influencing population growth or decline, and it can be used as a basis for a preliminary estimate of the direction of changes in population numbers. this can be a very useful feature if there is insufficient time to follow the change directly.

Age distributions can be measured and presented in several ways. The simplest measure for age is the raw age itself. This is probably the best scale to use in studies of a particular organism. However, it is difficult to compare age distributions of species with different life spans using raw ages, such as elephants and mice. Indeed, Stephen Stearns points out how the choice of measure for comparing the ages of different species may make a substantial difference in the analysis performed (Figure 5.8).[8] When comparing age distributions is important, it is often most useful to observe the maximum life span of the population and portray age as a percentage of that maximum life span. One can also determine the average age at death and present age groups in terms of deviation from this mean life span.

It is also meaningful to speak of *functional age*, where the life span is divided up into various functional units, such a prereoductive - reproductive - postreproductive, or egg - larva - nymph - adult. Individuals at different functional ages may appear quite similar, as in humans, or quite different, as in butterflies or other animals that undergo metamorphosis.

Figure 5.8
Illustration of the significance of age measures in comparing populations. *Tetraclita squamosa* is a subtidal barnacle that has a longer life span than the intertidal barnacle, *Chthamalus fissus*. Portraying age in absolute terms (a) points out the differences between the two species. Portraying age in relative terms (b) points out their great similarity. [Redrawn, with permission from Stearns, S. C. 1977. The evolution of life history traits. *Ann. Rev. Ecol. Syst.* 8: 145–171. Copyright © Annual Reviews, Inc.]

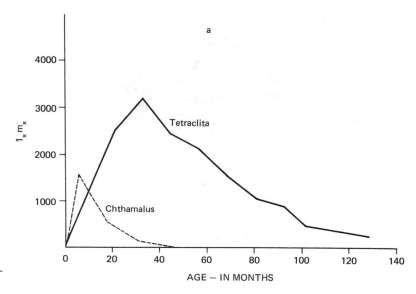

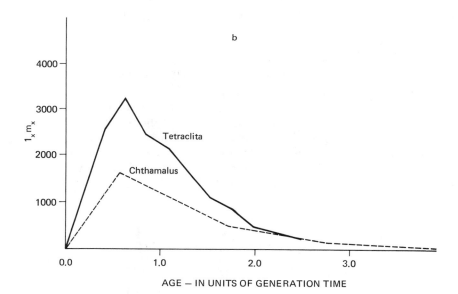

Functional age breakdowns can be powerful tools to point out the relative importance of specific environmental variables which act on different functional age groups. But they are virtually useless in studies involving gross age distribution, since different species may spend very different percentages of their lives in the various functional age groups. For instance, Figure 5.9 indicates tremendous differences between the life cycles of the two beetles, *Tenebroides* and *Trogoderma*, yet the percentage of life span spent in the various functional ages of *Tenebroides* is quite similar to that of humans. Even so, a functional age breakdown for a population can shed considerable light on what controls the species at different stages in its life cycle. This knowledge is often essential in designing a procedure to control a pest species on an economically important crop, and it is necessary to ascertain

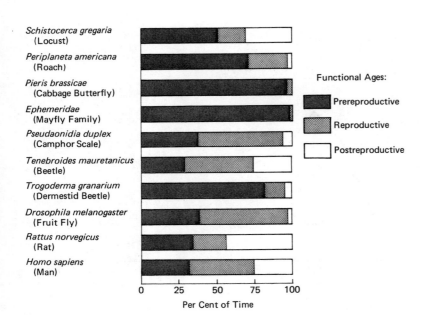

Schistocerca gregaria
(Locust)

Periplaneta americana
(Roach)

Pieris brassicae
(Cabbage Butterfly)

Ephemeridae
(Mayfly Family)

Pseudaonidia duplex
(Camphor Scale)

Tenebroides mauretanicus
(Beetle)

Trogoderma granarium
(Dermestid Beetle)

Drosophila melanogaster
(Fruit Fly)

Rattus norvegicus
(Rat)

Homo sapiens
(Man)

Functional Ages:

■ Prereproductive

▨ Reproductive

□ Postreproductive

0   25   50   75   100

Per Cent of Time

Figure 5.9
Percentage of time spent by various species of animals in various functional ages. [Redrawn, with permission, from W. C. Allee, A. E. Emerson, O. Park, T. Park, and K. P. Schmidt, *Principles of Animal Ecology.* Copyright © 1949: W. B. Saunders Company, Philadelphia.]

which stage in the pest's life cycle at what time of year is most vulnerable to attack by which means.[9]

Age Pyramids

The most common way of showing age distribution is the *age pyramid*. This is a vertical bar graph in which the number or proportion of individuals in various age ranges at any given time is shown, from youngest at the bottom of the graph to oldest at the top (see Figure 5.10). Age ranges may be presented in any of the formats discussed previously. An age pyramid is a "snapshot" of a living population at a specific moment; as the population changes through time, the form of the age pyramid does too. A progressive change of this sort is shown in Figure 5.11.

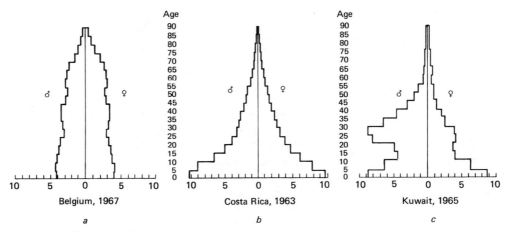

Belgium, 1967

a

Costa Rica, 1963

b

Kuwait, 1965

c

Figure 5.10
Age pyramids for various human populations, drawn with males and females shown separately as percentages of the total population: (a) Belgium, a stable population; (b) Costa Rica, a rapidly expanding population; (c) Kuwait, a population with extensive immigration, showing a highly skewed sex ratio. [Data from United Nations, 1970.]

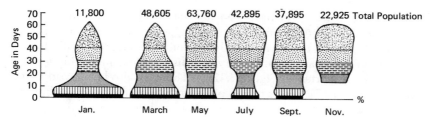

Figure 5.11
Progressive change in age distribution of a hive of honey bees over a single season. [Reprinted, with permission, from F. S. Bodenheimer, *Problems of Animal Ecology.* Copyright © 1938 by The Clarendon Press, Oxford.]

An age pyramid can provide a preliminary indication of whether a population is expanding, contracting, or stable. It is compared to the age pyramid of a stationary population (i.e., one that is neither expanding nor contracting and that will persist in the environment for a long period of time). The stationary age distribution reflects both birth and death, including the ages of both. The stationary age structure for an animal like the California condor, each pair of which produces a single egg every other year, might resemble Figure 5.12*a*; an animal like the oyster which produces millions of eggs, but whose offspring suffer high early mortality, might have a stationary age distribution like that in Figure 5.12*c*.

A broadened base in relation to the stationary age distribution indicates an increased number of juveniles and suggests that the population is expanding

Figure 5.12
Age structures of several different types of populations. The populations can be classified into those which produce virtually no excess young, such as the condor, those which produce a moderate number of young, such as man, or those which produce an exceedingly large number of young, such as the oyster.

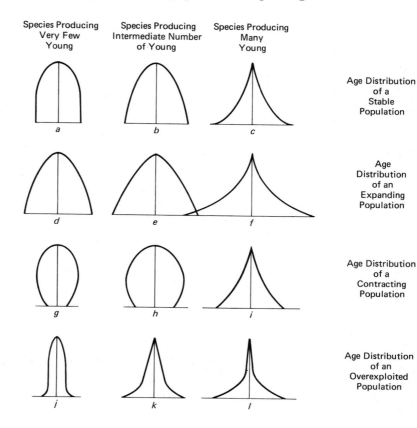

(Figure 5.12*d-f*). A narrower base than the stationary age distribution indicates a reduced number of juveniles and suggests that the population is contracting (Figure 5.12*g-i*). A heavily exploited population will show a much narrower top than the stationary age distribution, indicating greater than normal adult mortality (Figure 5.12*j-l*). This may also be a broad-based pyramid if any reproduction is taking place in the population at all, but the resemblance between the "expanding" pyramid and the "exploited" pyramid is only superficial. This final case is becoming increasingly common in commercially exploited populations of fish and wildlife.

Natality

The probability of giving birth or of dying varies with age, so that the vital rates of a population are a function of its age distribution. *Fertility* is the number of offspring produced by an individual. It is most often equal to zero until sexual maturity is reached, at which point it rises rapidly. In many species, the maximum fertility is reached in young adulthood, after which it falls off to zero if the organism lives long enough to reach the age of senescence. In other species, fertility is correlated with the size of the organism, and it increases throughout life. Figure 5.13 shows the relationship between fertility rates and age for different kinds of animals. The total number of young born in the population times the age-specific fertility rate for each age group (Table 5.2). Clearly, the age distribution of fertility may have a considerable effect on a population's birth rate, and the effect may be different for different populations.

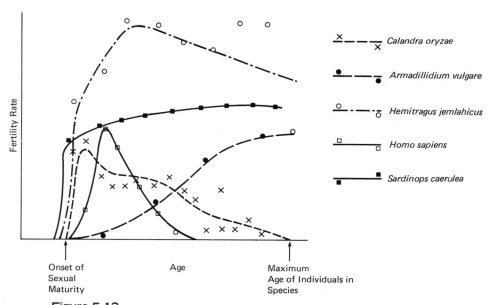

Figure 5.13
Smoothed curves showing the relationship between fertility and age for five different species: *Calandra oryzae*, the rice weevil (Birch, 1948), *Armadillidium vulgare,* a wood louse (Paris and Pitelka, 1962), *Hemitragus jemlahicus,* the Himmalayan thar, a goatlike ungulate (Caughley, 1970), *Homo sapiens,* man (Bogue, 1969) and the Pacific Sardine, *Sardinops caerulea* (Murphy, 1966) Actual fertility rates vary greatly for species; ordinate undefined.

Table 5.2	Age Range	Number in Population	Fertility Rate	Number Births
Reproduction in a Hypothetical Population of 100,000 Individuals Assuming Equal Sex Ratio	0–1	15,500	.00	0
	1–2	14,500	.20	1,450
	2–3	13,400	.50	3,350
	3–4	12,800	.40	2,560
	4–5	12,400	.30	1,860
	5–6	11,500	.20	1,150
	6–7	10,200	.10	510
	7–8	7,100	.00	0
	8+	2,600	.00	0
	Total	100,000		9,730

Survivorship

The equilibrium age structure of a population is the key to using age distribution statistics. This is derived not from studying the average number of living individuals in the population over a period of time, but rather from studying the ages at which they die. An animal dies only once, although it may be alive at many censuses. As a result, data derived from mortality do not change over the generation time. Also, because all animals die, the random fluctuations in population sizes and age distribution average out if complete mortality data are kept over several generations. Furthermore, mortality data for a given species may be roughly constant regardless of whether the population is expanding or contracting, unless the factors leading to the expansion or contraction do so by changing the relative probability of mortality at different ages. The graphic summary of a population's mortality data is termed a *survivorship curve*. It is a very powerful tool for dealing with the basic dynamics of population growth.

To construct a survivorship curve, a total population of individuals is considered at age 0 (birth). The total number of survivors is plotted at even increments of time, and a curve is drawn connecting the data. As an example, survivorship data for Dall mountain sheep in Denali National Park, Alaska, appear in Table 5.3. Figure 5.14 is a survivorship curve drawn from these data. Three classes of ideal curves can be recognized. No population, of course, has a curve that corresponds exactly to one of the ideal curves, but most can be observed to fit reasonably closely. The classes are as follows:

I. Probability of death increases with age.
II. Probability of death is constant with age.
III. Probability of death decreases with age.

The shape of the survivorship curve is related to the environmental stresses on the population. If there were no stresses at all, we might expect that an individual would live until it aged to a point that its body's physiology broke down or until it reached a physiological limit to its ability to survive. An early experiment with the fruit fly, *Drosophila melanogaster*, simulated the mortality of a population under conditions where the physiology of the individuals rather than the environment determined the survivorship.[10] Genetically homogeneous fruit flies were raised without food of any sort. As might be expected, they died rather quickly (70 hours, as opposed to the more normal 40 days). But because the flies were provided with no resources

Table 5.3
Life Table for the Dall Mountain Sheep, Based on Known Age of Death of 608 Sheep
Dying Before 1937 (Mean Length of Life, 7.09 Years)

Age Interval (Years)	Age as Percentage Deviation from Mean Length of Life	Number Dying in Age Interval Out of 1,000 Born	Number Surviving at Beginning of Age Interval Out of 1,000 Born	Mortality Rate per Thousand Alive at Beginning of Age Interval
0–½	−100.0	54	1,000	54.0
½–1	−93.0	145	946	153.0
1–2	−85.9	12	801	15.0
2–3	−71.8	13	789	16.5
3–4	−57.7	12	776	15.5
4–5	−43.5	30	764	39.3
5–6	−29.5	46	734	62.6
6–7	−15.4	48	688	69.9
7–8	−1.1	69	640	108.8
8–9	+13.0	132	571	231.0
9–10	+27.0	187	439	426.0
10–11	+41.0	156	252	619.0
11–12	+55.0	90	96	937.0
12–13	+69.0	3	6	500.0
13–14	+84.0	3	3	1,000.0

Reprinted, with permission, from E. S. Deevey, "Life Tables for Natural Populations of Animals,"
Quarterly Review of Biology, **22,** 283–314. Copyright 1947 by the Quarterly Review of Biology.

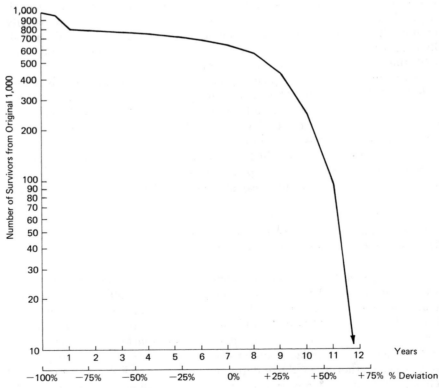

Figure 5.14
Survivorship curve for the Dall mountain sheep, drawn from the data in Table 5.3. Age is
expressed both in years and in percentage deviation from mean life span.

Figure 5.15
Idealized classes of survivorship curve, plotted with respect to percentage of life span. Curve 0 is the theoretical curve in the lack of environmental stress; curves I-III are as defined in the text.

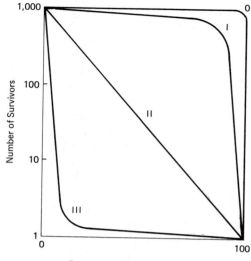

for their survival, there could be no intraspecific competition for resources, and no fly could gain a survival advantage over another through chance occurrences such as getting first crack at food. Their length of life was determined solely by the genetically based (and therefore uniform) physiological makeup of the population. The survivorship curve for these flies can be taken as an analog for survivorship of a population under conditions in which no abiotic environmental factor is limiting and no biotic factor has a negative effect.

The survivorship curves for actual populations under real conditions can then be compared to it (Figure 5.15). The species characterized by a class I curve under natural conditions tend to be settled. They are long-lived, as a rule, and they expend relatively little effort on reproduction. Examples include humans and most large vertebrates. Most other vertebrates and some invertebrates have Class II curves. They lose more of their numbers to predation, and so must expend relatively more effort in reproduction. They are likely to be shorter lived than those with Class I curves, and their numbers can be expected to fluctuate more. Opportunists tend to have Class III curves. These are the organisms that produce great numbers of larvae, most of which die while very young. Most plants also show this type of survivorship.

An Example of Population Regulation in a Real Community

Probably the best way to synthesize the materials of the last two chapters is to give a concrete example of an actual population, showing how various control mechanisms conspire to regulate its numbers, and delineating its interactions with different types of populations. We can show, by example, the extraordinary complexity of real populations, the ways different control mechanisms are brought to bear at different times, and the (small) degree to which any regulatory mechanism can be considered "the" population control device.

Very few natural populations have been studied so thoroughly that their regulatory mechanisms are well understood. One which has is *Cardiaspina albitextura* an insect of the psyllid family (jumping plant lice) which feeds on trees of the species *Eucalyptus blakelyi* in the region of the Australian Capital Territory.[11] Its population is typically stable at quite low densities, but it also can experience outbreaks in which the population gets so large that it seriously defoliates—and may even kill—the trees upon which it feeds. It is associated with many other species (Figure 5.16), some of which feed upon it during different parts of its life cycle, and some of which feed upon its predators. Some of the interactions between the species depend on the density of *C. albitextura*, while others seem independent of psyllid density. The triggering mechanisms for certain phases of the population dynamics of the species are biological, and others are dependent on features of the weather.

The most significant predators of *C. albitextura* are two encyrtid "wasps," *Psyllaephagus gemitus* and *P. xenus*. These are *parasitoids*: Their adults are free-flying insects, but their larvae are parasitic on *Cardiaspina*, which they kill. They are always active in the community, regardless of the *Cardiaspina* density, and they may infect up to 75 to 85% of the nymphs that are not killed by other means. The two species lay their eggs during different stages of the psyllid life cycle, with *P. gemitus* attacking the early nymphal stages and *P. xenus* attacking the later nymphal stages. The actual percentage of *C. albitextura* destroyed by the parasitoids depends on environmental conditions and does not seem to have any direct correlation to psyllid density. The parasitoids are unable to regulate the *C. albitextura* population by themselves, but they have an important role in the overall regulatory mechanism simply because of the large fraction of nymphs that they consume.

A third parasitoid species, *Psyllaephagus discretus*, does not normally feed on *C. albitextura* when its density is low, but it does attack when its density is higher. Thus, *P. discretus* is a density-dependent factor which can participate in a negative feedback regulatory mechanism.

Several species of birds use *C. albitextura* as a primary food supply As long as psyllid numbers are relatively low, bird predation efficiency rises as psyllid density increases. Thus the birds too are density-dependent regulators. However, the density-dependent negative feedback mechanism is not complete. The birds feed only on relatively mature nymphs and adults and have no effect on juveniles.

Under normal conditions, feedback-controlled predation by birds, encyrtids, and certain other predators (e.g. ants) operated to maintain *C. albitextura* at a relatively uniform low density, below outbreak proportions. This evidently held true for most of the period preceding the clearing of the southeast Australian tablelands by Europeans. However, efficient operation of these control mechanisms require that the stages in the life cycles of the various predators and parasitoids be synchronized with those of *Cardiaspina*. For instance, *Psyllaephagus gemitus* cannot reproduce if young *C. albitextura* nymphs are not readily available when adult female *P. gemitus* are searching for reproductive sites. Other species are no less restrictive in their requirements.

Of course, this would pose no problem if all ages of all species were present at all times. But not all stages in the life cycles of all species can survive the winter in southeastern Australia, and those stages that are present at winter's end are those that can survive the random variation in weather

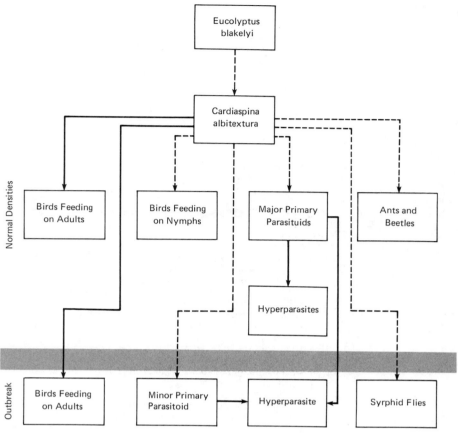

Figure 5.16
Organisms involved in the regulation of population size of *Cardiaspina albitextura* in the Australian Capital Territory. Arrows point in the direction of energy flow; dashed lines depict density-independent interactions; solid lines depict density-dependent interactions. Organisms above shaded line are involved at all densities of *Cardiaspina albitextura;* those below shaded line are part of the complex only at high densities of *C. albitextura.*

conditions. The ability of the parasitoids to control psyllid numbers may be substantially reduced when temperatures during one summer are above normal, followed by below-normal summer temperatures the next year. The efficiency of predation by *P. gemitus* and *P. xenus* is thus not only independent of psyllid density, but also directly related to weather conditions.[12] In addition, *P. discretus* and the birds are relatively inefficient consumers of *C. albitextura*, and they cannot keep the *C. albitextura* population under control unless a large proportion of the psyllid population is destroyed by the *P. gemitus* and *P. xenus*. If random climatic fluctuation leads to a decreased level of predation by these wasps, the rate of psyllid reproduction may be able to escape the control of the feedback mechanism, and the population may begin to increase to outbreak levels.

When the population escapes from negative feedback control, several more species become significant in regulating *C. albitextura*. Three are hyperparasites, or parasitoids of the parasitoids. The others are predaceous

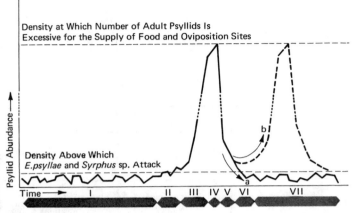

Figure 5.17
Summary of the population changes in *Cardiaspina albitextura* in the Australian Capital Territory over an outbreak cycle. Stages within the cycle are indicated by roman numerals. Control mechanisms during each stage are as follows:

"I. Psyllid numbers either stabilized or restricted to slow increase by the combined action of Predation by birds, ants, encyrtid parasites, etc., and The prevailing weather, especially temperature conditions;

II. Stabilizing process fails, e.g. because unusually low temperatures reduce percentage parasitism. Psyllid numbers rise to a level at which the hyperparasite *E. psyllae* begins to destroy primary parasites.

III. Psyllid numbers increase rapidly because of decrease in percentage destruction by birds, ants, and encyrtid parasites.

IV. Environmental opposition to population growth increases again because of increasing damage to foliage by psyllid nymphs.

V. Damage to foliage very severe. Psyllid numbers decrease greatly because of density-induced reduction in number of offspring per female, and excessive offspring for the number of favorable feeding sites available on foliage.

VI. Psyllid numbers further reduced to an extent, depending on predation intensity by birds, ants, encyrtid parasites, etc., which increases with decrease in psyllid numbers, and shedding of infested leaves by the host plant.

VII. If psyllid numbers are reduced below the level at which the hyperparasite *E. psyllae* operates, they are likely to remain low for some years. If not reduced to this extent, numbers tend to increase rapidly to the level at which the available amount of foliage again becomes limiting."

[Reprinted, with permission, from L. R. Clark, "The Population Dynamics of *Cardiaspina albitextura* (Psyllidae)," *Australian Journal of Zoology*, **12**, (1964), 349–361.]

flies of the genus *Syrphus*. Two hyperparasite species, *Psyllaephagus clarus* and *P. faustus*, normally feed on *Psyllaephagus gemitus*, *P. xenus*, and *P. discretus*. They can be found at all densities of *C. albitextura* (and hence of primary parasites), and the percentage destruction of primary parasitoids by hyperparasites is roughly independent of the density either of *C. albitextura* or of the primary parasitoids.

The parasitoids also become more common as psyllid densities rise to outbreak levels, and a third hyperparasite, *Echthroplexis psyllae*, enters the picture. The combined predation of *P. clarus*, *P. faustus*, and *E. psyllae* is density-dependent, and their ability to decimate the parasitoid populations increases with increasing parasitoid density. The upshot is that the parasitoid species that are needed to control *C. albitextura* are virtually wiped out by the hyperparasites during periods of psyllid outbreak. This allows the growth rate of *Cardiaspina* to increase even more. The decrease in the wasp predation is offset to some extent by predation on *C. albitextura* by *Syrphus* larvae, but *Syrphus* is much less efficient at controlling psyllid populations during outbreaks, since they are not numerous enough to be really effective except in situations where bird density is unusually high.[13] This is especially true in areas of human activity, where the habitat suitable for birds has been reduced by clearing the tablelands.

The net result is that the *Cardiaspina* population can rise virtually unchecked, once it has entered an outbreak phase, until it becomes so large that it seriously overexploits its food supply, causing massive defoliation and die-backs of branches. Only when this happens do late-nymphal *C. albitextura* begin to starve. Because there are fewer adequate oviposition sites, egg production decreases substantially. More nymphs die as they are shed by the millions along with dead infested leaves. The result is a rapid crash in the population of *C. albitextura* to a much lower level. The precise level is determined mainly by the intensity of parasitism and predation and by the extent to which psyllids are killed through leaf shedding.

If the population crash carries the levels of *P. gemitus* and *P. xenus* below the level at which they are attacked by *Echthroplexis psyllae*, feedback control can be reestablished over *C. albitextura*. However, if the crash ends with *E. psyllae* still being important, the most probable course is another outbreak and extensive defoliation.

The progression of a single outbreak of *Cardiaspina albitextura* is summarized graphically in Figure 5.17. The factors involved in controlling this species are complex, and different ones act at different times within the outbreak. It is hard to say how other species differ from *C. albitextura*, since millions of species inhabit the biosphere. But it is quite likely that *C. albitextura* is like many organisms in its population control mechanisms and that the tremendously complex system of interactions among populations is typical of ecosystems.

Notes

[1] Bogue, D. J., 1969. *Principles of Demography*. New York: John Wiley & Sons, Inc.

[2] Another way of expressing population growth is based on the discrete growth (compound interest) formula. The continuous growth equation presented here is somewhat easier to comprehend, however, for purposes of this discussion.

[3] MacArthur, R. H., and Wilson, E. O., 1967. *The Theory of Island Biogeography*. Princeton, N.J.: Princeton University Press; Wilson, E. O., and Bossert, W. H., 1971. *A Primer of Population Biology*. Stamford, Conn.: Sinauer Associates, Inc.; Stearns, S. C., 1977. The evolution of life history traits: A critique of the theory and a review of the data. *Ann. Rev. Ecol. Syst.*, **8**, 145–171; Green, R. F., 1980. A note on K-selection. *Am. Nat.*, **116**, 291–296.

[4] Stearns, S. C., 1977. The evolution of life history traits: a critique of the theory and a review of the data. *Ann. Rev. Ecol. Syst.*, **8**, 145–171.

[5] Huffaker, C. B., 1958. Experimental studies on predation: Dispersion factors and predator-prey oscillation. *Hilgardia*, **27**, 343–383.

[6] Pimentel, D., 1961. Species diversity and insect population outbreaks. *Ann. Entomol. Soc. Amer.*, **54**, 76–86; Pimentel, D., 1961. Animal population regulation by the genetic feed-back mechanism. *Am. Nat.*, **95**, 65–79.

[7] Wiens, J. R., 1977. On competition and variable environments. *Am. Sci.*, **65**, 590–597.

[8] Stearns, S. C., 1977. The evolution of life history traits: A critique of the theory and a review of the data. *Ann. Rev. Ecol. Syst.*, **8**, 145–171.

⁹ Clapham, W. B., Jr., 1980. Environmental problems, development, and agricultural production systems. *Environ. Conserv.*, **7**, 145–152; Clapham, W. B., Jr., 1980. Egyptian cotton leafworm: Integrated control and the agricultural production system. *Agric. and Environ.*, **5**, 201–211.

¹⁰ Pearl, R., 1928. *The Rate of Living*. London: University of London Press, Ltd.

¹¹ Clark, L. R., 1962. The general biology of *Cardiaspina albitextura* (Psyllidae) and its abundance in relation to weather and parasitism. *Aust. Jour. Zool.*, **10**, 537–586; Clark, L. R., 1963. On the density and distribution of newly established nymphs of *Cardiaspina albitextura* (Psyllidae) at times of high abundance. *Proc. Linn. Soc. New South Wales*, **88**, 67–73; Clark, L. R., 1964. The intensity of parasite attack in relation to the abundance of *Cardiaspina albitextura* (Psyllidae). *Aust. Jour. Zool.*, **12**, 150–173; Clark, L. R., 1964. Predation by birds in relation to the population density of *Cardiaspina albitextura* (Psyllidae). *Aust. Jour. Zool.*, **12**, 349–361; Clark, L. R., 1964. The population dynamics of *Cardiaspina albitextura* (Psyllidae). *Aust. Jour. Zool.*, **12**, 362–380; Clark, L. R., Geier, P. W., Hughes, R. D., and Morris, R. F., 1967. *The Ecology of Insect Populations in Theory and Practice*. London: Methuen & Co., Ltd.

¹² Clark, L. R., 1962. The general biology of *Cardiaspina albitextura* (Psyllidae) and its abundance in relation to weather and parasitism. *Aust. Jour. Zool.*, **10**, 537–586.

¹³ Clark, L. R., 1964. Predation by birds in relation to the population density of *Cardiaspina albitextura* (Psyllidae). *Aust. Jour. Zool.*, **12**, 349–361.

Niche and Community

6

POPULATIONS EXIST in communities, and it should be clear by now that many of the most significant factors that influence populations lie in the structure of the community. The notion of the community has been implicit in much of our discussion so far, even when it has not been mentioned explicitly. The community can be defined as the assemblage of populations living in an ecosystem. Its significance lies not so much in the populations themselves, however, as in the interactions between them and in the characteristics these interactions confer on ecosystems.

We have seen how individual populations interact strongly with one another, both as food supplies and as regulatory agents. But populations also interact with other members of their communities in more subtle ways that have little direct role in the regulation of the specific groups involved but that are fundamental to maintaining the community as a whole, and hence to the existence of any species in the community. For instance, white-tailed deer in a Pennsylvania forest do not interact directly with nitrogen bacteria. They belong to different food chains; both have their own population dynamics, and they are regulated by different aspects of the physical or biotic environment. Without nitrogen bacteria, however, nitrogen could not cycle completely, and deer could not exist (nor any other form of life, for that matter). Conversely, without members of the grazing food chain such as deer, nitrogen bacteria would have no nitrogenous wastes to feed upon. The interdependence among species goes far beyond the easily defined pairwise interactions discussed in Chapter 4. The very existence of any population is due to the fact that other populations also exist which make available energy and nutrients, regulate population size, recycle wastes, and buffer the community against the vagaries of the physical environment. Populations exist, in fact, because they are parts of communities.

Community Structure

The dynamic interactions among populations vary continuously in time and space. But communities have a structure at all times and in all situations that is reflected in the roles played by their constituent populations, their ranges and the types of areas they inhabit, the diversity of species in the community and the spectrum of interactions between them, and the precise flow patterns of energy and nutrients through the community.

Why do you find the species you do in the places they are? This seemingly naive question goes to the heart of the concept of community structure. Populations are characterized by *place* and *function*. They occur where they do because of the functions they discharge. The place of any population is its *habitat*. Its function is its *ecological niche*. It is sometimes meaningful to speak of a population's combined place and function as its *ecotope*.[1]

These concepts can be seen from the perspective of the population or of the community. The distinction may be important in understanding the phenomena, but it is not essential. Indeed, a population's niche in the community does not depend on whether a population biologist or a community ecologist (or anybody else) is studying it. Habitat and niche both refer to attributes of a particular population in a particular community, and they belong to both levels of organization.

The Ecological Niche

The key concept in community ecology is the ecological niche. It expresses a population's role in a community and in the ecosystem. It is based on the bonds between the population and all other populations with which it interacts, directly and indirectly, as well as the bonds between the population and its inorganic environment.

One can visualize these bonds in terms of the tolerance range—optimum model which is so useful for environmental resources. Each resource in the ecosystem is characterized by a feasible gradient, within which populations may or may not be able to survive. Such a gradient could be described as in Figure 3.1 (page 47). A similar gradient could be shown in a second resource, along with the population's responses (Figure 6.1). If there were

Figure 6.1
Schematic response of a population to variation of two resources, showing responses to variation in each resource and to both together.

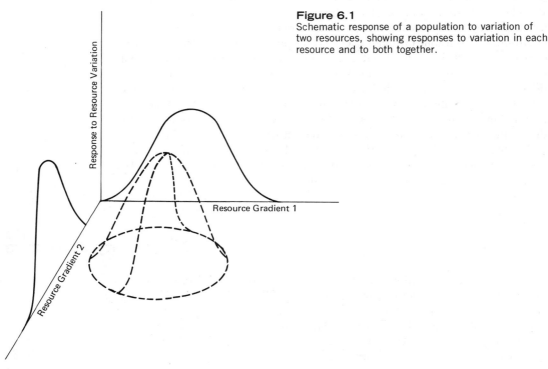

more than two dimensions to a piece of paper, we could go on and draw more resource gradients and the appropriate population responses.

At a higher level of abstraction, we could also express the population's interactions with other species on this same diagram, until we had completed a graphic representation of the population's resource requirements and interactions with the suite of environmental factors present in the ecosystem. It would be a complicated drawing, to be sure, but it would describe the responses of the population to any value for any variable in the suite of factors considered. We could then modify the drawing so that it also included the population's effects on all of these factors. The result of this exercise would indeed be a comprehensive picture of the population's role in the community, its niche.

It is obviously impossible to draw a picture of a niche using a paper and pen. Even using mathematical formulations, people have not been successful at doing more than beginning to explore the complexity of natural ecosystems. The niche is a construct—a concept that describes the role of each species in the ecosystems of which it is a member.

The ecological niche is a property of each population in each community. Understanding the niche provides an explanation for the population's structural, physical, and behavioral adaptations. But it is never easy to define a population's niche. Many of the factors included in it are more significant than others, or their influence on the population's adaptations or behavior is more obvious than others. These may mask the significance of other factors. In the same way, the contribution of any single factor to a population's niche may change as the complex of factors constituting the environment changes. Change of this sort can happen in many ways, as the seasons change, as communities develop in a given area (ecological succession; see page 146), or as evolution at the species level allows major changes in the structure of communities (community evolution; see page 156).

The ecological niche is a property of the community, although it is defined in terms of populations. Understanding the niches of the populations composing a community provides insight into the interactions binding the populations. It is these interactions that make a community more than the sum of its constituent parts and that confer on the community whatever order it contains.

It is always dangerous to talk about "order" in natural systems, since we are accustomed to thinking in terms of ordered human systems, which were constructed for a purpose by known individuals. The order in natural ecosystems is built by natural selection, acting over a long period of time. Species have evolved to a point where they can survive in an ecosystem. The previous abstract description ("drawing") of the niche can be regarded as a description of a multidimensional "resource space" in which the species can survive. In this particular case, the difference between calling a significant variable a "resource" or an "interacting population" is semantic. The population in question responds to it by surviving or not and by reproducing (if it survives) more or less effectively. Furthermore, the population's response is dynamic. The acts of being alive and doing the things one must do in order to keep alive provide stimuli to other populations and affect their responses. Not every organism can do everything equally well. A degree of specialization is imposed on species by the simple existence of the myriad of signals in the ecosystem: chemical, visual, energy, behavioral. Those who can gain

Figure 6.2
Ecological equivalents:
(a) The eastern meadow-
lark (left) is an Ameri-
can Icterid, whereas
the yellow-throated
longclaw (right) is an
African Motacillid. (b)
the dovekie (above) is
a member of the same
order as the gulls,
whereas the Magellan
diving petrel (below)
belongs to the same
order as the albatross.
[Redrawn, with permis-
sion, from J. Fisher
and R. T. Peterson,
The World of Birds.
London: Aldus Books,
Ltd., 1964. Copyright
ⓒ 1964 by Roger Tory
Peterson and used with
his permission.]

their resource needs do so, and those who do not, do not. In essence, evolution divides up the multidimensional resource space into a set of niches. It is not imposed from outside for a specific purpose, but it is definitely built up.

One indication of the role of evolution in dividing the available "niche space" is the phenomenon of *ecological equivalence*. Different communities in ecosystems in different places characterized by similar environments are often exceedingly similar in their structure. Populations occupying similar niches in the two communities may have the same suites of adaptations, even though they are totally unrelated. For instance, the succulent desert plants of the southwestern United States are cacti; deserts of southern Africa contain plants of the spurge family that are virtually identical in structure. Figure 6.2 shows two pairs of bird species, each of which is totally unrelated to the other, yet which are very similar in appearance and basic adaptations.

Ecological equivalents provide a certain amount of insight into the nature of the niche. The adaptations of organisms are due to selection for survival in the here-and-now. For cacti and spurges to have such similar forms, or for dovekies be so similar to Magellan diving petrels suggests that the here-and-now was rather similar in each pair of cases. We can measure the details of the physical environment and determine how similar it is. In fact, the deserts of southwestern North America and southern Africa are similar in many ways, as are the environments of the birds. But evolution responds also (and perhaps more strongly) to biological signals, and for these to be similar enough to bring about ecological equivalence implies a similar structure to the communities themselves.

Habitat

The niche describes the function of the population in the ecosystem. The habitat describes its geography. The two concepts are closely related, but they also complement each other. The niche deals with the factors with which a population interacts: the signals generated by the abiotic environment or by other populations. The habitat deals with the landscape pattern, the environment as a set of geographic patches, gradients, or what-have-you. In some ways, the habitat is more concrete than the niche. Because it is a geographically based, one can point to the habitat of a species and say, "Look, there it is." This is impossible with the niche.

The habitat of a species may be broad, covering thousands of square kilometers, or it may be very restricted. The American bison, for example, once ranged over the entire center of the North American continent. Certain species of leaf miner, on the other hand, live only in the upper photosynthetic cell layer of the leaves of certain species of plants, with other species living in the lower cell layer. There is an important distinction between the *range* of a species and its habitat. The former refers to its broad geographic extent: central North America for the bison or an oak-hickory forest for a particular kind of leaf miner. The latter refers to the location within the range where all of the environmental factors required for its presence are realized. The difference may not mean much for the bison, but it is quite significant for the leaf miner.

The needs of specialized populations such as the leaf miners are so consistent that they clearly mark different habitats. Subdivisions of the environment on this scale are commonly called *microhabitats*. The leaves constitute a microhabitat for leaf miners within the total forest, and the different cell layers of the leaf constitute different microhabitats within the leaf for different species of leaf miner. Conditions within the leaves are quite different from the general conditions in the forest. The specific environmental variables in the microhabitat of a population can be referred to as the microenvironment or the microclimate.

Niche Space and the Divisions Between Niches

The significance of the niche concept is intuitively obvious. But it remains a difficult notion to define adequately,[2] and it is even more difficult to work with analytically. Understanding communities in terms of niches requires defining and working with the niches of tens, hundreds, or even thousands of different species. The question of why we find species where they are is easy to ask but extraordinarily difficult to answer.

A good place to begin is with the notion of niche space developed on page 132. This was originally based on an important paper published in 1957 by G. Evelyn Hutchinson[3] and built upon many times.[4] The population's responses to the gradients in various resources are noted and the overall niche space for the population is described. This can be done easily for two resource gradients and with difficulty for three. For more than that, we must rely on mathematical abstraction.

Likewise, the responses of two species that use the same resources can be shown easily for one or two resources, with difficulty for three, and mathematically thereafter. The comparison between the one-resource case and the two-resource case is instructive, however. Figure 6.3 shows the response of two populations, A and B, to variation in a resource gradient of some sort. Their peak response is different, but they show considerable overlap. When the responses of A and B are plotted with respect to two different resources

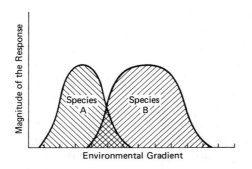

Figure 6.3
Diagram of one facet of the niches of
two populations: responses of the populations to an environmental gradient.

(Figure 6.4), the degree of overlap is smaller. In fact, the only way it could
not be smaller is for the requirements of both species to be identical for one
of the two resources.

How likely is niche overlap? Figure 6.5 shows the food preferences of
several herbivores in a semiarid part of Colorado. Six of the animals are
grasshoppers, one is the Mormon cricket, and one is the Richardson ground
squirrel. There is tremendous overlap in what the animals *can* use for food,
but the amounts of each plant species actually consumed by each herbivore
is quite different. The squirrel is no more different from the grasshoppers in
its food consumption patterns than they are from each other or from the
Mormon cricket.

Most populations in a natural ecosystem show varying degrees of niche
overlap. After all, all species require certain resources, and one can expect a
certain amount of competition for them *a priori*. The problem is to determine when niche overlap is such that coexistence is possible, and, conversely, when it is such that certain populations cannot survive the stresses
of interspecific competition. There may also be times when new populations

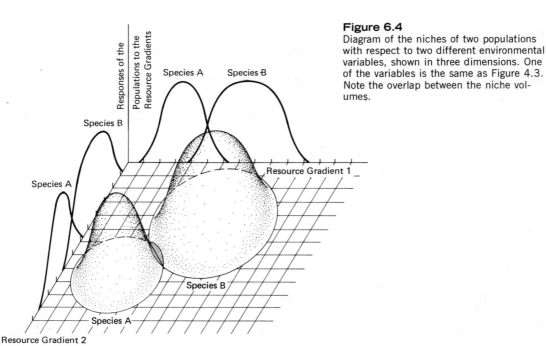

Figure 6.4
Diagram of the niches of two populations
with respect to two different environmental
variables, shown in three dimensions. One
of the variables is the same as Figure 4.3.
Note the overlap between the niche volumes.

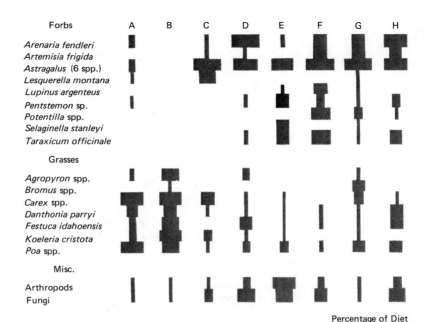

Figure 6.5
Average percentage dry weights of certain food items in the diets of eight herbivores at Prairie Divide, Colorado: A–*Xanthippus corallipes,* a grasshopper; B–*Aeropedellus clavatus,* a grasshopper; C–*Circotettix rabula,* a grasshopper; D–*Melanoplus alpinus,* a grasshopper; E–*Anabrus simplex,* the mormon cricket; F–*Melanoplus bruneri,* a grasshopper; G–*Citellus richardsonii,* the Richardson ground squirrel; H–*Melanoplus infantilis,* a grasshopper. [Data from Hansen and Ueckert, 1970.]

Percentage of Diet by Dry Weight

0–2%
2–4%
4–8%
8–16%
16–32%
32–64%

can enter a community, forcing a readjustment of its niche structure but allowing preexisting populations to survive.

Everything else being equal, natural selection within a given community will tend to lead to increasing niche separation, so that niche overlap is minimized, and competition for essential resources is minimized. During the development of a community, two populations with significant niche overlap may become established in an area (Figure 6.6*a*). Because the genetic makeup of natural populations is fairly heterogeneous, we can think of certain individuals from each population as lying in the area of overlap, while other members of the same populations do not.

Consider, for example, two species of plants that are able to grow in saline water, such as in a salt marsh. Species A can grow successfully if the salt content of the soil solution is 0 to 2%, and species B can grow successfully if the salinity of the soil solution is 1 to 3%. Those individuals from both populations that are physiologically best adapted to survive under salinities of 1 to 2% compete directly with one another. Because the individuals of species A that favor a salinity lower that 1% and those of species B that favor a salinity greater than 2 % do not compete directly, they are likely to survive in relatively greater numbers than their fellows in the area of overlap, and pass more genes on to their posterity (Figure 6.6*b*). The reduction in the area of overlap is equivalent to separation of the niches. There are two implications of niche separation that have far-reaching effects on the structure of a community. First, the remaining niches are both smaller and more specialized than they were to begin with; second, the population sizes of one or both of

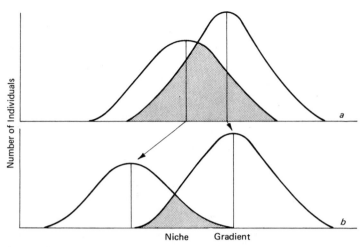

Figure 6.6
Diagram showing the process of niche divergence during the development of a community: (a) two species normally have very similar niches; if they are placed in direct competition with one another, the individuals in the shaded area are at a selective disadvantage with respect to those individuals within the populations which are not in the area of niche overlap; (b) competition between species leads to a shift in the niches of the two species, as individuals outside of the area of overlap are favored by natural selection; both species become more specialized, and the degree of niche overlap is substantially reduced. [Redrawn, with permission, from R. H. Whittaker, *Communities and Ecosystems*. Copyright © 1970 by Macmillan Publishing Co., Inc., New York.]

the species are likely to be smaller than they would have been in the absence of the other.

Species Abundance and Diversity of the Community

Different localities can support different kinds of communities. Some can support thousands of species, while others can support only a small number. The number of species in the community is termed its *richness*. Few aspects of biogeography are clearer than the increasing richness of communities as one travels from a depauperate environment such as a desert or arctic eco-

Table 6.1
Relationship Between Richness of Community and Harshness of Environment As Shown by Numbers of Species in Representative Taxa from Florida to the Canadian Arctic

Taxon Approximate Latitude	Florida (27°N)	Approximate Number of Species		Baffin Island (70°N)
		Massachusetts (42°N)	Labrador (54°N)	
Beetles	4,000	2,000	169	90
Land Snails	250	100	25	0
Intertidal Mollusks	425	175	60	°
Reptiles	107	21	5	0
Amphibia	50	21	17	0
Fresh-Water Fishes	°	75	20	1
Coastal Marine Fishes	650	225	75	°
Flowering Plants	2,500	1,650	390	218
Ferns and Club Mosses	°	70	31	11

* Data lacking.
Modified, with permission, from G. L. Clarke, *Elements of Ecology*, rev. printing. Copyright 1954 by John Wiley & Sons, Inc., New York.

system to a rich one, such as a rainforest or coral reef (Table 6.1). Even so, richness is not a particularly useful concept by itself, since it may be difficult or meaningless to compare communities on the basis of richness alone. For instance, a community spread over a large area can be expected to be richer than a geographically restricted community, if for no other reason than that the statistical probability of finding a rare species in a larger area is greater than it is in a small area. Also, two communities have quite different structures if 30 populations in the first each account for 2 to 5% of the total, but 2 or 3 species in the second account for 95% of all the individuals in the community.

The degree to which different populations approach equality in their relative abundance is termed the *equitability* of the community.[5] In no community are all species equally abundant: Some are always more common than others. The distribution of relative abundance from the most common population to the least common is a significant aspect of the structure of a community, and it seems to vary in different types of communities. One of the simplest descriptions of relative abundance of species in a community is the so-called log-normal distribution.[6] Despite its simplicity, there is compelling evidence that it provides a very useful general picture of species abundance for most natural communities.[7] This graphic description can also provide useful insight into other aspects of community structure, such as diversity and dominance.

A typical instance of log-normal distribution of species abundance is depicted in Figure 6.7a. For simplicity's sake, only plants are shown. The horizontal scale describes "importance": This can be measured in terms of numerical counts, density, biomass, role in mineral cycling, or some other measure of the population's importance or role in the community. The height of the histogram represents the number of species whose abundance falls within the range plotted on the horizontal scale. The relative abundance or importance of species in a community with log-normal distribution can also be represented by ranking the populations from most abundant (rank 1) to least abundant (rank n) and plotting each species on a semilogrithmic graph; this yields a figure such as Figure 6.7b.

DIVERSITY. Diversity is a measure of the integration of a biological community that includes both the number of constituent populations and their relative abundances. If the distribution of species abundance in several communities is the same, diversity is directly related to the number of constituent populations. However, two equally rich communities may have different diversities if the relative abundance of species in one is more equitable than that in the other. The community with the greater equitability would be considered to have the higher diversity. This is reasonable, considering that the community would *appear* to be richer, since the casual observer would perceive more different species on first observation.

The importance value curve shown in Figure 6.7b shows both aspects of diversity. The richness component is indicated by the number of populations plotted, and the equitability component is shown by the slope and shape of the curve. Highly equitable communities have a curve that is horizontal or convex upward; those of low equitability have curves that are more concave upward. One can also utilize several mathematical diversity indexes which need not be discussed in detail in this book but which have proven useful.[8]

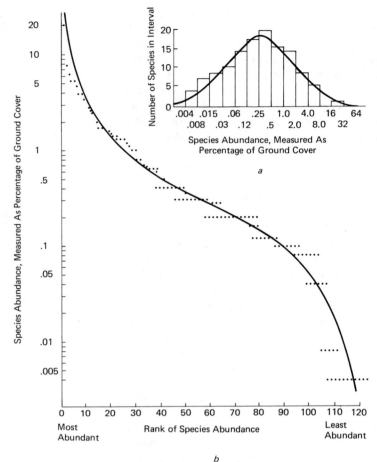

Figure 6.7
Lognormal distribution of species abundances of plant species in the Santa Catalina Mountains, Arizona. Abundance is measured in percentage of ground cover by the species: (a) histogram showing number of species in intervals of abundance, with fitted lognormal curve (modified, with permission, from R. H. Whittaker, "Dominance and Diversity in Land Plant Communities," *Science*, **147** [January 15, 1965], 250–260. Copyright © 1965 by the American Association for the Advancement of Science; (b) importance value curve showing abundance of plant species ranked in order of ground cover [drawn from data supplied by Dr. R. H. Whittaker].

No single measure of community diversity is appropriate for all studies. But many measures can provide exceedingly useful information about the structure of communities.

DOMINANCE. Some species in any community have a much greater influence on what goes on in the community than others. This influence varies from place to place, but it can be visualized in terms of the changes that would result if the species were removed altogether. It is probable, for instance, that removal of the lady's slipper, *Cypripedium reginae*, from a hardwood forest in Massachusetts would make little difference to the economy of the forest as a whole, if only because most herbivores feeding on it could transfer to one of the myriad other species of small plants present. Removing beech (*Fagus grandifolia*) or sugar maple (*Acer saccharum*) from the same forest would remove a large proportion of the biomass of the community and would cause great changes in the forest. Clearly, beech and maple have a much larger role in the forest than the lady's slipper, if for no other reason than that they constitute most of the forest's biomass. When a population such as that of the beech or maple is thought to exert an unusually strong influence on the other populations in the community, it is referred to as a *dominant* species.

It is easier to define the dominance concept than to deal with it practically. "Influence" is not an objectively measurable parameter, because it refers to things that are often difficult or even impossible to quantify. Numerical abundance is inadequate, since small organisms may be more abundant than large organisms without having as much impact on the total community. Biomass is generally a better measure of a population's presence in the community. Gross or net productivity reflects a species' ability to exploit its environment and is a more successful way to compare organisms of widely varying size. Beech and maple in the Massachusetts forest qualify under both of these criteria. But certain populations with relatively low biomass or productivity may occupy strategic niches with respect to nutrient cycling or other key ecosystem functions. An example of this is *Cactoblastis cactorum*, following its introduction into Australia (page 93).

It is difficult or impossible to apply the dominance concept across trophic levels, and the characteristics of a dominant autotroph may be very different from those of a dominant herbivore or carnivore. Within a single community, it might be most feasible to measure the importance of the plants by ground cover, of herbivores by biomass, and of carnivores by their net production. We may even speak of a dominant plant, a dominant herbivore, and a dominant carnivore within a single community, as though there were no interaction between them. Identifying particular species as dominant is not always an objective matter. Like the limiting factor concept, it is a simplifying abstraction that provides the observer with a degree of insight into the structure of the community. It is often a useful notion, but the reader must not forget that it is a simplifying model that may mean different things to different people, and that a community may or may not have a population (or several populations) that can be considered dominant in any meaningful fashion.

Patterns of Community Distribution in Space

Communities are never spatially uniform. The ranges of specific populations within the community are different, and patterns of community structure and related phenomena may vary from place to place. Most aspects of community distribution are ultimately under the control of fundamental abiotic factors such as temperature, rainfall, and salinity. Gross variation patterns are controlled by regional differences in the environment, and finer patterns are controlled by local heterogeneity in the ecosystem.

Diversity, Heterogeneity, and Equilibrium

No ecosystem is homogeneous with respect to either its abiotic components or its biotic makeup, with the possible exception of waters of the open ocean. Heterogeneity involves major features such as topography, as well as environmental grain (page 115). The resources available to species in different subenvironments may be quite different. To a degree, at least, the distribution of species reflects these resource differences. Some species are restricted, and others range unimpeded throughout the area.

Patterns may be either horizontal or vertical, and they may reflect abiotic or biotic factors. For instance, the flowers and other small herbs in a hardwood forest are most common in areas where the canopy is somewhat broken and light can reach the ground. Most stream organisms are restricted to certain environments within the stream, such as riffles, deep pools, or shallow backwaters. Many bird species are very particular about the level within the forest at which they reside; the upper and lower parts of the canopy and the shrub or small tree layers may be inhabited by different species of birds, even though any of them could easily fly anywhere in the forest. The vertical distribution of many lake-dwelling organisms is strongly regulated by light intensity, and it changes from day to night.

Different factors control ecosystem heterogeneity in different places. But the overall effect can determine the diversity of the community as a whole, especially when rare species are involved.[9] This is particularly true in complex communities where different species may be specialized for different parts of the broad spectrum of microhabitat differences.

Major environmental features can be regarded as "given." They are constant except over geological time. But microhabitat and grain are not at all constant, and it is significant whether they are in equilibrium with the other factors of the environment. Equilibrium, in this case, implies a balance among the forces emanating from the various sectors of the ecosystem, so that the configuration of microhabitats would not change if there were no perturbations from outside the system. This is an important matter, since much of the current theoretical work in community ecology assumes that ecosystems are in (or at least very near) equilibrium. There is evidence that this is not, in fact, the case.[10] Nonequilibrium ecosystems are actively changing, and the mix of signals in a changing system is very different from that in an equilibrium system. Nonequilibrium systems are also likely to be more heterogeneous environments than equilibrium systems. These factors affect both the diversity of communities and their potential stability.

Ecosystems with nearly constant environmental conditions tend to have communities with high species diversity, everything else being equal. The high predictability of the fundamental environmental factors allows a high degree of specialization within the community. Specialists can be more efficient than generalists in meeting their resource requirements, if the resource base is sufficient to allow niche diversification to take place. A corollary of specialization is a large number of populations in the community, with dominance playing a minor role. Stress exists in predictable environments, but to a degree different from other ecosystems. Because species are more highly specialized, a relatively small shift in environmental conditions may constitute an acute stress.

Many ecosystems are characterized by variable environments. They are not predictable, and fluctuations in one or more fundamental environmental factors may be significant. Survival in such an ecosystem requires a population to be adapted to a wide range of conditions. Narrow niches suited to specialists do not exist long enough to allow specialization to be built into the community by natural selection. Populations in unpredictable environments must be generalists.

As if to complicate matters, environmental uniformity and predictability are not the only determinants of community diversity. Overall productivity determines the ecosystem's resource base and is thus an important factor in determining the number of populations that can exist in a given area. In a rainforest or a coral reef, for instance, productivity is very high, and a great many species can be supported. In the depths of the ocean, however, productivity is so low that few individuals of any species can be supported there. Thus the diversity of the deep ocean bottom cannot approach that of a coral reef or a rainforest, even though its environmental constancy is virtually unmatched in any other ecosystem. Even so, the diversity of deep ocean bottom communities is much higher than that of more shallow marine bottom communities.[11] In the same way, tundras and deserts have low-constancy environments. Their productivity and diversity are also low. But they are easily matched in variability by the estuary, the brackish-water zone between many rivers and the sea (page 198). But an estuary is a very fertile ecosystem, and it is as productive as a rainforest or a coral reef. The estuarine community is much less diverse than that of either the river or the sea (see Figure 7.29), but it is also richer than that of the tundra.

Community Gradients and Boundaries

It is often difficult or impossible to determine where one community ends and the next begins. Many communities, in fact, grade continuously into each other with no noticeable boundary. For example, the differences between a hardwood forest of the northeastern United States and a Canadian spruce forest are obvious. The diversity of the former is much greater, and the species constituting the communities are largely different. They are obviously two very different sorts of communities. However, if one travels from a location in which the hardwood forest is well developed, such as central Pennsylvania, to a spruce forest in northern Ontario, it is impossible to locate the boundary between them.

Intergradation between different communities that are connected by an environmental gradient is quite common. The gradient can be temperature (as in the forests mentioned earlier), rainfall, depth in a body of water, or some other factor. Figure 6.8 shows the distribution of plant populations along two moisture gradients in mountainous areas of the western United States. The composition of the vegetation at the end points of the continuum is quite different, although some species are found throughout the gradient. The range of any one plant species has no fundamental relationship to that of any other. Even in much more complex communities than those shown here, where two species may be unusually similar, the vast majority of species have overlapping ranges that are determined by their own requirements and characteristics. Only rarely, in instances of environmental discontinuity or disturbance such as fire or human interference, is a sharp boundary between communities likely or even possible.

There are some relatively sharp boundaries in nature, however, where the gradient separating communities is steep and is best thought of as a *transition zone* separating them. In such situations, it is possible to observe the types of interactions that occur between communities, as well as certain properties of the transition zone itself. Similar phenomena probably occur with the much more widespread small-scale discontinuities resulting from ecosystem heterogeneity, but they cannot be observed on that scale.

Transition zones between communities are quite variable. They may be only a few meters wide, or they may be much wider. Figure 6.9 diagrams the species distribution across a typical transition zone, as between a prairie and a forest. The community on either side of the zone may have a typical structure, but the transition zone is strikingly different. Individuals of all species from both communities can migrate constantly into this zone, and it

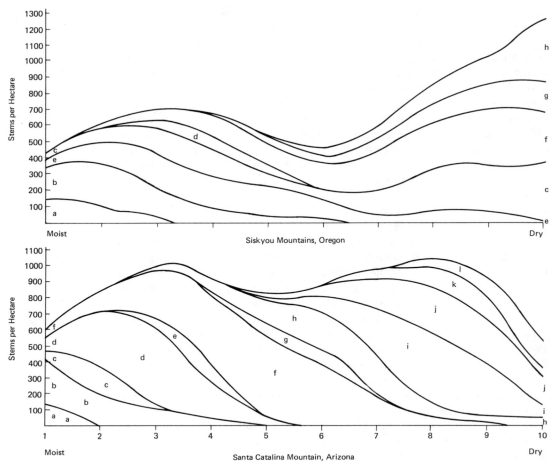

Figure 6.8
Change in composition of vegetation along moisture gradient: (*a*) Siskyou Mountains, Oregon; several species can be found from one end of the gradient to the other, but the composition of the vegetation changes markedly; (*b*) Santa Catalina Mountains, Arizona; only one species is found along the entire gradient, such that the composition of the vegetation changes qualitatively as well as quantitatively. Data from Whittaker, 1967.

really does not matter whether long-term environmental conditions would normally preclude a given species from becoming established there, since it can repopulate the area as soon as limiting conditions have passed. The result is a zone that has a higher diversity than either of the main communities.

In addition to species from each of the main communities, the transition zone may contain certain populations specialized to the transition zone itself. This potential for the transition zone to act as a habitat for species found in neither major community is termed the *edge effect*. A common example is those species of owl that live in forests near grasslands. They depend on forest trees for nesting, and they do their hunting in the grassland, where they depend on field rodents for food. In man-made communities such as agricultural fields, the transition zone between the field and the forest may act as a refuge for species formerly found in the plowed area, as well as for other plants such as weeds. Transition zones of this type are also the prime habitat of many species of insect, game bird, and mammal.

Figure 6.9
Diagram of the species distribution across a transition zone between two communities, labeled A and B: (a) distribution of the species characteristic of each of the major communities and of the transition zone itself; (b) total diversity of the communities across the transition zone.

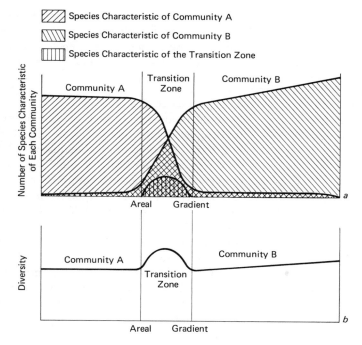

No boundary can be drawn in nature that is not crossed by something significant. The shoreline, which would seem the most definitive of community boundaries, is regularly crossed by adult frogs, some insects, otters, beavers, snakes, shore birds, and muskrats, to name only a few. Several plants are found on both sides of the shoreline. All amphibians and many insects spend their larval period in the water and their adulthood on land. Many large ungulates such as moose and deer have their summer range in one area and their winter range in another. Habitual passage from one community to another is the norm for many species in all parts of the world.

Even when organisms in adjacent ecosystems seldom interact, stress conditions such as forest fire, unusually deep snow, and certain kinds of intermittent pollution can cause animals to move temporarily across the boundary into a refuge from which they can repopulate their own habitat after the stress conditions have passed.

Even major barriers to interaction such as mountains or oceans are not absolute. Individual organisms or small colonies can be carried great distances and become permanently established in places where they do not "belong." Often, as in the case of the invasion of the Great Lakes by the sea lamprey and the alewife, or of Australia by the rabbit, the new population can wreak havoc within its new environment. Actually, this is not all that uncommon when viewed in the perspective of geologic time, but the rate has been markedly increased by human activity.

Alteration of Community Patterns in Time

Communities change. Their change can be understood on several levels. The simplest is the growth, interaction, and death of individual organisms as they pass through their life cycles, affected by the cycles of seasons and other natural phenomena. But other levels of community change act over longer time spans and account for much larger changes in community composition and structure. These include *ecological succession*, the reaction of communities to progressive development in regional environmental factors, and *community evolution*.

Ecological Succession

An abandoned field or a severely burned-over forest begins as little more than a plot of bare ground. Some plants take root and grow there, establishing a simple community. Gradually, other species become prominent and may displace some of the original populations. The changes in the structure and composition of the community are often rapid at first, slowing gradually until a point of dynamic equilibrium is reached, and the community is more or less stable (Figure 6.10). The first species to become established are the *pioneers*; the community that ends the sequence is the *climax*. This process of continuous community change is ecological succession; it is characteristic of all ecosystems. It may proceed from an initial state in which a preexisting community has been removed (as through farming or fire), or it may proceed from an initial state in which no previous community has existed (as on bare volcanic rock or sand dunes). The development of the ecosystem is continuous, although it is often useful to visualize the progression in terms of gradational stages.

No successional model applies to all situations. Perhaps the most useful overview of the principles of succession is by Joseph Connell and Ralph Slatyer.[12] They define three different models, of which the first is the standard textbook model and the other two are often overlooked.

The *facilitation* model for succession occurs when the entry of late-successional species into the ecosystem depends on earlier forms "paving the way."

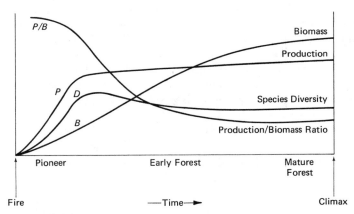

Figure 6.10

Progression of community parameters through succession following a fire in a generalized terrestrial environment. Primary production rises rapidly during pioneer stages, then more slowly. Biomass rises relatively slowly at the beginning, when the community is composed primarily of herbs and shrubs, then increases more rapidly as these give way to trees, and finally reaches its maximum in the climax forest. Species diversity rises to a maximum as the early forest trees invade the area, then drops slightly as pioneer species are eliminated. The ratio of primary production to biomass reaches its maximum during the early stages of succession, when production is largely from plants of relatively small biomass, then drops as trees with high biomass account for an increasing proportion of primary production. [Redrawn, with permission, from R. H. Whittaker. *Communities and Ecosystems.* Copyright © 1970 by Macmillan Publishing Co., Inc., New York.]

Climax ecosystems are very different from pioneers; their communities could not have survived under pioneer conditions, and their structure is much better integrated than that of the pioneers. Despite the fact that this is the textbook example of succession, the evidence for it is fairly sparse. It certainly occurs in many *primary successions*, where the succession begins with bare rock, sand, or a similar environment. The *tolerance* model is characterized by a change in the strategies of different organisms for acquiring resources. Late-successional species may be able to tolerate lower levels of resources than earlier ones, and the structure of climax communities is different from that of the pioneers, but the survival of late-successional organisms does not depend on the prior existence of the earlier stages, and in fact many late-successional organisms may be among the first invaders of the disturbed habitat. In the *inhibition* model, the first-comers are dominant for an extended period of time and inhibit further change. They are replaced only when they die or are damaged in some way, releasing resources to others. This is evidently rather common in nature.

FACILITATION SUCCESSION. Facilitation is the basis of the most classic examples of ecological succession. It exemplifies the dynamic linkage of the ecosystem's living and nonliving facets as few other phenomena can. Changes in abiotic factors are imposed by the developing community. But as the abiotic components change, their influences on the community change, and the community itself changes in response. Change does not stop until there is a balance among the forces impinging on the ecosystem that allows an equilibrium to become established. This dynamic interaction among all facets of the ecosystem is shown in Figure 6.11.

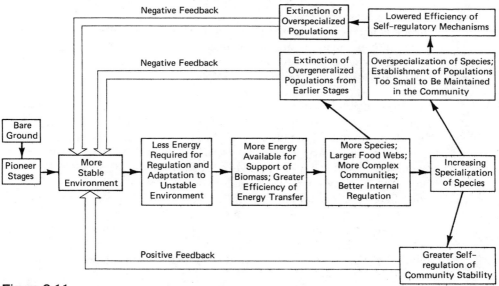

Figure 6.11
Schematic representation of the progressive changes in communities during ecological succession as a series of feedback loops. During the early stages, the lower positive loop is dominant, leading to a self-sustained increase in community diversity and stability. During the later stages of succession, the introduction of more specialized species via the lower loop is counteracted by the loss of overgeneralized or overspecialized populations by the upper negative feedback loop. The climax community is maintained by a dynamic equilibrium between the two loops. [Based on Connell and Orias, 1964.]

The pioneer assemblage that colonizes bare rock or open sand inhabits a harsh environment. Temperature variations are commonly wider than in a mature forest, nutrients are less available, and fewer species have the adaptations required for survival. Pioneer species are generalists with wide niches, able to withstand wide fluctuations in abiotic factors. The primary productivity and biomass of the pioneer community tend to be low, but the ratio of primary production to biomass is high (Figure 6.10). This ratio is a measure of the inefficiency of energy transfer through the community. Its high level during pioneer stages indicates either that it takes a great deal of energy to support a given amount of biomass, or that generalist pioneer populations are not very efficient in disposing of their energy throughout the community. Diversity of the pioneer community is low, so that food webs are poorly developed and interspecific interactions are minimal. Nutrient movement into or out of the system proceeds readily. The pioneer community is not a highly organized, stable entity. It exists, however, because the populations that constitute it are adapted to endure the conditions found early in succession.

The changes brought about by the developing community enable other organisms to enter the ecosystem. The basic characteristics of the environment become less limiting. The action of plant roots fractures rock or stabilizes shifting sand, and the community captures nutrients and other resources imported from adjacent ecosystems. Productivity in the climax is high, and it may even be at a maximum for the entire succession, although the productivity peak may also be reached fairly early, followed by a gradual drop to the steady-state level. Biomass is at its highest during the climax. The ratio of primary production to biomass is low, indicating that a relatively large biomass can be maintained on a given input of energy; this indicates that the climax represents a peak in the efficiency of energy disposition throughout the community. Diversity is much higher than in the pioneer stage, and populations are specialized to relatively narrow niches. Food webs are typically well developed, and the import or export of nutrients from the ecosystem is relatively unimportant.

TOLERANCE SUCCESSION. The differences between tolerance and facilitation in succession are illustrated in Figure 6.12, which contrasts the three models outlined by Connell and Slatyer. Pioneers in both models are those species that can survive, but they are a small subset of the total in the facilitation model. They can include a broad spectrum of species in the tolerance model, including those characteristic of practically any stage in succession. Late-successional species are not precluded from the pioneer stage, they just don't characterize it. They tolerate the existence of the pioneer species and grow slowly in the background. The pioneer species do have an influence on the abiotic environment, and they tend not to replace themselves, but they do not make critical improvements in the environment that allow the later species to enter.

The best example of this type is old field succession, in which a forest reclaims an abandoned agricultural field. Unless the soil has been totally destroyed by bad farming practices, propagules of trees, grasses, and other plants can get started in the soil. The most prominent plants at first are those that mature quickly: the grasses and broadleafed herbs. The juvenile trees are there too, but they are not as obvious. The shade beneath the pioneers

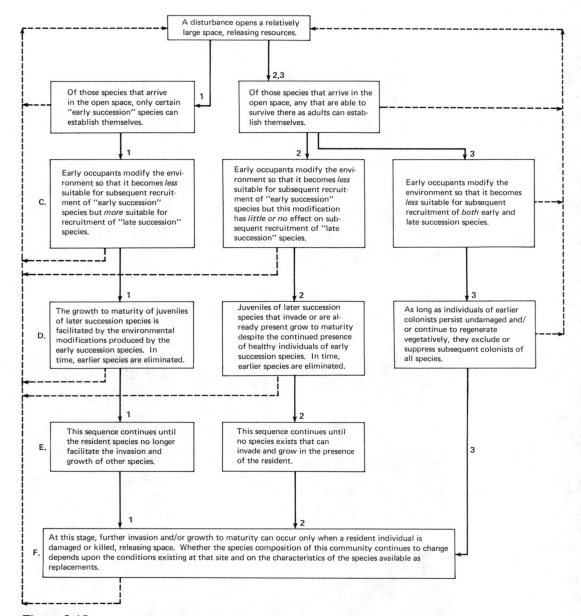

Figure 6.12
Three models of the mechanisms producing the sequence of species in succession. Solid lines indicate successional sequence. Dashed lines represent breakdowns and return to original state.
1. Facilitation.
2. Tolerance.
3. Inhibition.
[Redrawn, with permission, from Connell, J. H. and Slatyer, R. V. 1977. Mechanisms of succession in natural communities and their role in community stabilization and organization. *Am. Nat.* 111: 1119–1144. Copyright © University of Chicago Press.]

a

b

c

d

Figure 6.13
Stages in dune succession in the Indiana Dunes National Lakeshore, at the south end of Lake Michigan
    a. Unvegetated dune.
    b. Dune stabilized with low vegetation.
    c. Dune covered with well developed vegetation. Soil contains some organic material.
    d. Mature dune with well developed forest. Soil contains considerable organic material.
[Photos courtesy U. S. National Park Service, by (a&c) Richard Frear,
(b) M. W. Williams, and (d) William Dutton.]

inhibits the reproduction of most pioneer species, but the plants characteristic of later stages tolerate it. For this reason, they gradually become more prominent, and the earlier stages are forced out of the community as the later stages mature.

It is significant that the trends of productivity, biomass, and other indications of ecosystem "improvement" used to characterize facilitation succession also apply to tolerance succession, but for a different (and simpler) reason. The later-successional communities in a typical terrestrial succession are forests, which are larger and more efficient at cycling energy and nutrients than communities composed mainly of grasses and shrubs.

INHIBITION SUCCESSION. In the third model of succession, the early-successional species inhibit further change. Colonists interfere with the establishment of new species. When an individual dies, it does not have to be replaced by a species with different life-history characteristics. It may be replaced by the same species, in which case there will be no change in the community. It may also be replaced by a longer lived species that characterizes later successional phases. This is unlikely in one sense, in that these species are suppressed by the pioneers. But it does happen, and there will be a gradual change in the community away from the pioneer configuration if even a small percentage of short-lived pioneers is replaced by long-lived later-stage species. The rate of change is determined by the tolerance of the inhibited species to their suppression and by the relative life span of the species involved.

There are many examples of this kind of succession. Plants can inhibit others by allelopathic secretions (page 90) that inhibit later-stage populations chemically, or they can form such a dense shade that later-stage plants simply cannot grow. For example, shrubs in utility rights-of-way can suppress tree growth for periods of over 40 years.[13] Experiments on growth of marine animals on tiles suggests that the first sedentary marine invertebrates to colonize the tile are able to retain their hold until they die and slough off.[14] Even in developing forests with considerable light penetration to the ground, young trees do not fare well in competition with older trees. There is more light than in a more mature forest, so that light is not limiting, but root competition for water and nutrient minerals is still more important.

AUTOGENIC AND ALLOGENIC SUCCESSION. The community is the primary source for changes in the physical environment and in the community itself in *autogenic succession*. This is typical of many successional sequences, including most terrestrial ecosystems. But this is not universal. In a lake, for instance, the input of nutrient materials from outside is typically more important in controlling the changes in the physical environment, while development of community structure are largely a function of the physical environment. This is *allogenic succession*. Let us examine the course of succession in an example of each type.

Succession in an Area of Sand Dunes. A classic example of succession in a terrestrial ecosystem is the progression from bare dune to a forest dominated by beech and sugar maple in the Indiana Dunes at the southern end of Lake Michigan (Figure 6.13). These dunes are composed of sand that has been winnowed out of the beach deposits by the wind and built up into high dunes along the lakeshore in Indiana and Michigan. The sand shifts constantly in the wind, and bare dunes are very unstable substrates for life. In addition, the sand is very permeable, so that water passes easily through it to a deep water table. The dunes are dry and desertlike—in fact cacti are not uncommon in some areas. A few species, mainly grasses, are able to colonize the bare, shifting dunes. But the grasses stabilize the sand, allowing entry into the community for plants that can withstand low water levels but cannot tolerate the shifting sands.

As individuals of these species die, some of their organic matter is incorporated into the sand as humus. This has two effects: It binds sediment particles, further stabilizing the dune, and it binds water. As more water is bound in the upper layer of the dune, new species can enter the community, and productivity rises accordingly. As productivity rises, so does the biomass of the plants and animals, as well as the amount of detritus added to the sand. The detritus level, production, community biomass, and community diversity spiral upward until the dune is covered by a forest of oaks and blueberries, and the soil is a brown humic sand rather than a clean white sand. Eventually, water is replaced by nutrients as the primary limiting factor. But as humus builds up, the nutrient levels follow (Figure 6.14). Finally, the soil is sufficiently rich to support the beech—sugar maple climax forest that is typical of this area.

This is the classic example of facilitation in autogenic succession. The description by Jerry Olson shows clearly how development of the community brings about changes in the abiotic environment and enables further changes in the community.[15]

Succession in a Small Pond or Bog. An equally classic, but very different pattern of succession is the allogenic succession from a pond or bog to a forest. Sediments containing nutrient materials, organic molecules, and mineral grains are washed in from outside. The community responds directly to these imports, and the changes that occur are largely controlled by the imported materials. The net result is typically a late-successional community of much lower diversity than it was originally. This type of development is characteristic of lake eutrophication; it will be discussed in greater detail in Chapter 7.

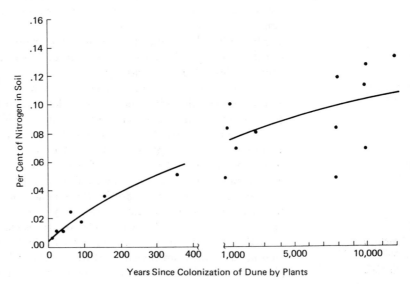

Figure 6.14
Change in nitrogen in the top 10 cm. of soil in dunes of several different ages at the south end of Lake Michigan. This change is correlated with the succession of plants and animals on the Lake Michigan dunes, and it is similar to the change in several other abiotic factors occurring at the same time. [Adapted from Olson, 1958.]

Per Cent of Nitrogen in Soil

Years Since Colonization of Dune by Plants

A recently formed pond may be fairly deep, with a low nutrient content and very clear water. It is fed by rivers, which bring in sediments and dissolved nutrients. As sediments build up on the bottom and nutrient materials accumulate in the water, the pond becomes more fertile and shallower, with productivity increasing as a result. As productivity increases, the structure and composition of the pond community change, often showing a marked drop in diversity.

As the pond becomes shallower, rooted plants become more common near the shore. These plants trap sediment particles washed in from around the pond, and the pond may begin to fill in from the shore. A floating mat of *Sphagnum* moss and sedges can become established around the edge of the pond, or bog, as it is now more likely to be called. This mat is even more efficient as a sediment trap that the rooted plants, and a soil may become established on top of the floating mat. The bog continues to fill from the top as the mat builds up in thickness, from the sides as the mat increases its extent out into the bog, and from the bottom, as sediment and organic material accumulate beneath the mat. Eventually, the bog is filled, and terrestrial succession can continue to yield a forest climax.

Succession and Equilibrium

One of the most significant and enduring myths in ecology is that the climax stage of ecological succession represents an equilibrium, that the end point of succession is an ecosystem with a stable configuration. This implies that the climax is both identifiable as an end point and stable. Neither of these is necessarily the case.

It was once believed that all ecosystems in a region would show succession to a single climax type controlled by the regional climate, given sufficient time, regardless of the original surface upon which the succession started or the composition of any of the intermediate stages. The prophet of this notion was the late F. E. Clements.[16] He established the idea of the climax as an immutable end point to a degree that has been very hard to shake. It became clear by the mid-1950s that topography, grain, soil variation stemming from differences in parent rocks, and similar environmental variables can lead to significant variations in the climax.[17] It is becoming clear at this time that it is extraordinarily difficult to put one's finger on a stable climax in the field, and that a climax is little more than an ecosystem that doesn't change very much during a course of study.[18]

Reaching a climax stage suggests constancy: The young of a particular species replace the adults when they die, so that there is no change in the gross composition of the ecosystem in time. What this means is easiest to visualize with plants. When a canopy tree of a particular species dies, what species will usually replace it? The understory is seldom uniform (although it sometimes is), and the winner of competition for the space in the canopy may be one of several species (see Figure 4.12).

Three alternatives are possible. If the adaptations of the tree are such that the probability of self-replacement is very high, then the community will not change. This is often the case, and many trees can resprout from their roots even after the crown and stem have died. These include many climax trees, such as the American beech, and the southern beech of Australia. But the adaptations of many species neither favor nor interfere with self-replacement. A canopy tree may be replaced by a conspecific tree or another species. It probably depends most on the composition of the understory, which may reflect conditions at the current time or at some previous time when the understory was becoming established. Finally, there are trees that tend not to replace themselves without a certain amount of disturbance. These include the redwoods[19], and the cedars of Lebanon.[20] Both are commonly regarded as climax communities, but the fact that the juveniles of neither plant can thrive beneath adults indicates that the climax is a moving one, not a stable equilibrium. Indeed, A. S. Watt documented a process by which the vegetation of an area in England showed a constant cycle among different communities. Each was successively replaced by another, until the first returned.[21]

In summary, it is not at all clear that climaxes ever represent equilibrium communities. To quote Connell and Slatyer, "We have found no example of a community of sexually reproducing individuals in which it has been demonstrated that the average species composition has reached a steady-state equilibrium. Until this is demonstrated, we conclude that, in general, succession never stops."[22]

This should not really bother us too much, although it is fuzzier than we might like. All it means is that ecosystems respond to a remarkably complex set of signals, some of which are current (e.g., the information and the concrete signals), and some of which stem from past conditions (e.g., the composition of a forest understory or the organisms available to colonize a particular area following clearing).

Succession Considering Changes in Regional Climate

Regional climates are not constant. Over the last 14,000 years, for example, the climate of the northern United States and southern Canada has varied from very cold (following the retreat of the continental ice sheet) to somewhat warmer than the present, then cooling slightly to yield the climate of the present day. Given this, one might expect to see major changes in the communities occupying any given area during this period. This has indeed been the case.

The history of vegetation in many areas can be reconstructed from the fossil record of pollen and spores produced by plants and deposited in lake, bog, or marine sediments. The postglacial records of pollen and spores are quite extensive, and we know a great deal about the progression of vegetation over the last 14,000 years, as communities responded to the progressive changes in climate.[23] Figure 6.15 records the distribution of pollen of several plant groups recovered from a core of lake sediment from southern Connecticut. If we assume that pollen production per plant and the probability of fossilization of the pollen from a given species have not changed significantly throughout the period (and these seem to be valid assumptions), the number of grains preserved at any given level can be used to estimate the composition of the forest present at the time that sediment horizon was deposited.

Thus, southern Connecticut was covered first by tundra following the retreat of the glacier, which was later intermixed with spruce woodland. About 9,500 years ago, the area was covered by a mixed forest dominated by pine, birch, alder, and oak, with the abundance of pine slowly decreasing and that of oak slowly increasing until some 7,000 years ago, when a mixed hardwood

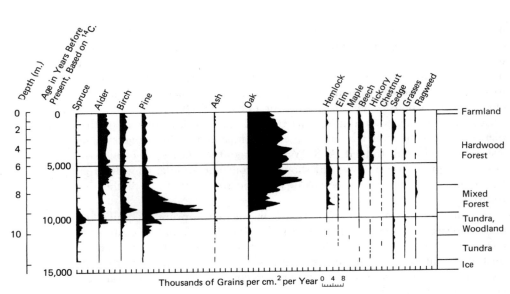

Figure 6.15
Accumulation rates for most important types of pollen from Rogers Lake, Connecticut. The record extends from the final retreat of the ice sheet in the area until the present time. [Redrawn, with permission, from M. B. Davis. "Pollen Accumulation Rate at Rogers Lake, Connecticut during Late- and Postglacial Time," *Review of Palaeobotany and Palynology*, **2** (1967): 219–230. Copyright © 1967 by the Elsevier Publishing Company, Amsterdam, Holland.]

forest that would last for some 6,700 years was established. The hardwood forest has declined as agriculture has become widespread during the last 300 years.

The compositions of the plant assemblages indicated by Figure 6.15 are quite similar to those of presently existing communities ranging from northern Canada to New England.[24] This would suggest that communities respond to climatic change largely by modifying the ranges of entire community groups to conform to the changing distribution of climatic zones. Indeed, there is much about the responses of vegetation to postglacial climatic changes that can be explained in these terms. But these responses of communities to climatic shifts are no more deterministic than ecological succession. The dynamic interaction between populations that determines the nature of a community can bring about new community aggregations if natural selection so dictates. The vegetational change following the retreat of the glaciers illustrates demonstrable progression in the composition of specific communities in addition to progressive displacement of the ranges of basic community types.[25]

A final illustration of the dynamic interactions that characterize communities involves the evolution of community structure as species evolve in response to natural selection. Because evolution is exceedingly slow, we cannot observe community change at this level, but a few instances from the fossil record can show the process in action. One of the most instructive examples demonstrating evolution of the basic structure of the community concerns the development of a terrestrial community of a modern type some 250 million years ago.

Between the time when vertebrates (amphibians) first became able to lead a predominantly terrestrial existence some 350 million years ago and the establishment of an essentially modern type of food web some 100 million years later, the structure of the terrestrial community was decidedly different from what it is now. Development of the modern set of interactions among large vertebrates required not only a complete rearrangement of the niche structure of the community but also the evolution of new species that could fill the new niches.[26]

Attaining the adaptations needed for terrestrial life by the first amphibians did not in itself establish a land-based vertebrate community. These amphibians were carnivores, and the only animals inhabiting the land environment were arthropods (first scorpions, centipedes, millipedes, spiders, etc.; later insects). The early amphibians were heavy and slow. They were ill-equipped to capture quickly moving prey. The first communities inhabited by terrestrial vertebrates are best regarded as extensions of the aquatic communities. The land habit was an adaptation to improve the capabilities of predators whose prime food supply was aquatic invertebrates and fish (Figure 6.16a).

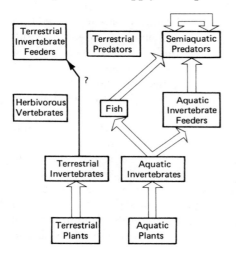

a

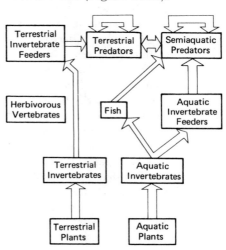

b

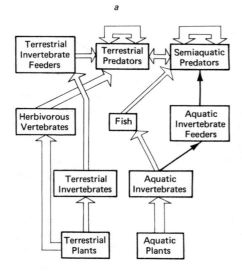

c

Figure 6.16
Stylized food webs at several stages of the development of the modern type of terrestrial vertebrate community: (a) immediately after the realization of the land habit by early amphibians; vertebrate food chains entirely water-based; 350 million years ago; (b) intermediate stage; vertebrate food chains largely water based; 300 million years ago; (c) modern type of land vertebrate community, realized by the development of terrestrial herbivorous vertebrates; food chains largely land-based; 250 million years ago. [Figures b and c modified, with permission, from E. C. Olson, "The Food Chain and the Origin of Mammals," *International Colloquium on the Evolution of Mammals*, Part 1, 97–116. Brussels: Koninklijke Vlaamse Academie voor Wetenschappen, Letteren en Schone Kunsten van Belgie, 1961.]

By some 300 million years ago, reptiles had evolved that could feed effectively on terrestrial invertebrates (Figure 6.16b). An entirely land-based community was theoretically possible in which all herbivore niches were assumed by invertebrates and some of the carnivore niches by vertebrates. However, the evidence we have suggests that most contemporary carnivorous vertebrates were unable as yet to realize an entirely terrestrial carnivore niche, so that the great majority of the energy flow through the community continued to pass through the aquatic route. The typical food chain to the highest terrestrial vertebrate carnivore was plant $\longrightarrow$ aquatic invertebrate $\longrightarrow$ aquatic-invertebrate-feeding vertebrate $\longrightarrow$ semiaquatic predator $\longrightarrow$ terrestrial predator.

By 250 million years ago terrestrial herbivorous vertebrates had evolved, and a fully terrestrial vertebrate community could come into being (Figure 6.16c). From this time onward, the basic structure of the terrestrial community was of an essentially modern sort, with all consumer trophic levels occupied by a wide range of animals, both vertebrate and invertebrate.

Such evolutionary changes in the structure of communities are caused by a large number of factors. One, of course, is the regional climate. It was becoming progressively drier during the period under consideration, and the perfection of land-based adaptations is a reasonable response to this sort of change. Indeed, many evolutionary changes in community structure can be explained on the basis of responses to major changes in the regional abiotic factors of the environment.[27]

But many instances of community evolution, including the one just outlined, are much broader, involving the expansion of the community's structure by organizing niches that had not previously existed in the community. The fact that this type of change can be demonstrated to have occurred in the past supports the view of ecosystems as dynamic and creative. This is true not only as an abstraction but also from a practical standpoint, considering our own species' ability to modify, at least temporarily, communities of which we are parts and to create niches which may or may not be stable in the long run.

Heterogeneity, Diversity, and Stability

Few issues in ecology have been the subject of more controversy than the relationship between stability and diversity of ecosystems. This is more than a theoretical argument. It has some immensely practical aspects. Specifically, people depend on ecosystems for certain products, including food, wood, textile fibers, rope fibers, fish, and so on. The resource and amenity values of ecosystems depend on their responses to stress. An agricultural system that collapses periodically cannot support an industrial civilization. It is no fun to go walking through the woods and spend the day dodging gypsy moth larvae that are in the middle of a major outbreak. Algal blooms in lakes used for drinking water can be bothersome, to say the least. For those of us who live on the shore of Lake Erie and drink its water, a blue-green algae bloom like the one shown in Figure 3.4 is not an abstraction; we taste the algae every August, and our bathrooms reek of them after a hot shower.

There is a widespread feeling that community diversity is related to ecosystem stability. After all, agricultural monocultures tend to be vulnerable to pest outbreaks, and this tendency can be reduced by intercropping several different cultivated plants. Islands are vulnerable to invasions by species from outside, and they suffer a much more rapid turnover of species than larger, more diverse mainland communities. The lower diversity communities of the Arctic and Antarctic fluctuate violently, while the diverse communities of tropical rainforests and coral reefs appear to be stable. Diversity has been interpreted as a way of buffering the community against change. It is an expression of W. Ross Ashby's "law of requisite variety," which states that maximum stabilization of a system requires a sufficient number of self-regulatory feedback mechanisms to compensate for the spectrum of environmental challenges presented to it.[28] A diverse community is characterized by a very high number of niche interactions, many of which are regulatory in nature. A simple community with many fewer regulatory interactions is less well buffered.

On the other hand, much of the theoretical work done since 1970 using mathematical models of community interactions suggests strongly that diversity is correlated with instability, not with stability. Many scientists, in fact, are convinced that the correlation of diversity and stability has been discredited. The most complete statement of this argument is a review by Daniel Goodman, which should be read by anybody seriously interested in the controversy.[29]

One of the difficulties in dealing with the correlation of stability and diversity is that "stability" does not always mean the same things to everybody involved in the discussion. It can mean the fluctuations in the abiotic environment (*environmental constancy*). It can mean the degree of oscillation of populations in the community (*community constancy*). It can mean the ability of the system to return to its equilibrium condition when perturbed by some stress condition (*resilience*) or the tendency of the system to remain in its equilibrium condition when stressed (*resistance*). Finally, there is the ability of the ecosystem to remain as a viable system despite stress, induced oscillations, and the by-products of being out of equilibrium for some period of time (*persistence*). In general, the theoretical study of stability has concentrated on resilience of very simple communities, whereas the practical considerations of community stability concern constancy and persistence. Models showing that higher diversity decreases resilience also show that it increases resistance.[30] It is not clear what this means to the ecosystem, but it can well imply increased constancy.

Heterogeneity and Diversity

We can agree on certain things. Some environments are much more diverse than others. Tropical rainforests and coral reefs, for example, are every textbook's example of diverse ecosystems, and with reason. In the same way, tundras and deserts are examples of impoverished ecosystems. It is also true that change is obvious in the communities of low diversity, while it is subdued in the diverse ecosystems. This is not to say that tropical communities do not have epidemics or that populations in diverse environments do not oscillate. They do. But the evidence we have suggests that oscillations do not characterize the diverse ecosystems in the same way they characterize the simple ones.

In any case, one must remember that all natural ecosystems have a certain amount of persistence. However adaptations influence community structure, natural communities that have persevered have staying power, oscillations or no. The practical problems come in when we increase the level of perturbations above their normal levels.

It is instructive to examine the responses of a community such as a tropical rainforest to stress. The community is highly buffered against change. As suggested by Ashby's law of requisite variety, the feedbacks conferred by the high diversity appear to buffer the forest and confer a high degree of resistance. If, however, the community is actually perturbed beyond its normal limits, the specialists may not be able to tolerate the perturbation, and the community may break down. This has happened to tropical forests both in central Africa and (even more so) in the Amazon. It is becoming abundantly clear that such communities are not resilient to major perturbations.[31] This is consonant with the notion that stability (resilience) decreases with increasing diversity, and also that stability (resistance) increases with increasing diversity. Neither addresses the question of persistence.

Compare this response with the response to stress by a simple community from an area of high environmental variability. The low diversity provides minimal buffering against the vagaries of environmental fluctuation, so that all (or at least most) populations are characterized by major oscillations in density. Resistance to change is low, but resilience must be high if the species are to survive at all.

Another influence on both diversity and stability is grain. Microhabitat variables can buffer a community from certain types of stress, and many microhabitat factors stem from the activities of organisms. For example, the floor of a conifer forest is protected from winds, evaporation, and temperature shifts, making it a much more predictable environment for organisms than the forest canopy or an open field adjacent to the forest (see Figure 8.19). They can also alter the nature of interspecific interactions. Spatial heterogeneity can allow diversity to develop even when more fundamental factors would not. We have already seen this with the crayfish *Orconectes virilis* and *O. immunis* in the Little Sioux River in Iowa and Minnesota (page 87).

Heterogeneity can provide refugia for populations that would otherwise become locally extinct. Refugia mean persistence for the populations and resilience for the community. We have seen this with the mites, *Eotetranychus sexmaculatus* and *Typhlodromus occidentalis* (page 92). It almost does not matter whether the increased diversity allowed by coarse grain confers additional stability (resistance) on the ecosystem as well. Grain is one of the key aspects of any ecosystem, and it is an important factor in all interpretations of stability.

Notes

[1] Whittaker, R. H., Levin, S. A., and Root, R. B., 1973. Niche, habitat, and ecotope. *Am. Nat.*, **107**, 321–339.

[2] MacArthur, R. H., 1968. The theory of the niche. Chap. 11 in Lewontin, R. C., ed. *Population Biology and Evolution*. Syracuse, N.Y.: Syracuse University Press, 159–176.

[3] Hutchinson, G. E., 1957. Concluding remarks. *Cold Spring Harbor Symposium Quant. Biol.*, **22**, 415–427.

[4] Maguire, B., Jr., 1973. Niche response structure and the analytical potentials of its relationship to the habitat. *Am. Nat.*, **107**, 213–246.

[5] Lloyd, M., and Ghelardi, R. J., 1964. A table for calculating the "equitability" component of species diversity. *Jour. Animal Ecol.*, **33**, 217–225.

[6] Preston, F. W., 1948. The commonness, and rarity, of species. *Ecology*, **29**, 254–283.

[7] Sugihara, G., 1980. Minimal community structure: An explanation of species abundance patterns. *Am. Nat.*, **116**, 770–787.

[8] Pielou, E. C., 1969. *An Introduction to Mathematical Ecology*. New York: John Wiley & Sons, Inc.; MacArthur, R. H., 1965. Patterns of species diversity. *Biol. Rev.*, **40**, 510–533; Lloyd, M., and Ghelardi, R. J., 1964. A table for calculating the "equitability" component of species diversity. *Jour. Animal Ecol.*, **33**, 217–225.

[9] Hairston, N. G., 1959. Species abundance and community organization. *Ecology*, **40**, 404–416.

[10] Connell, J. H., 1978. Diversity in tropical rain forests and coral reefs. *Science*, **199**, 1302–1310.

[11] Sanders, H. L., and Hessler, R. R., 1969. Ecology of the deep sea benthos. *Science*, **163**, 1419–1424.

[12] Connell, J. H., and Slatyer, R. O., 1977. Mechanisms of succession in natural communities and their role in community stability and organization. *Am. Nat.*, **111**, 1119–1144.

[13] Niering, W. A., and Goodwin, B. H., 1974. Creation of relatively stable shrublands with herbicides: Arresting "succession" on rights-of-way and pastureland. *Ecology*, **55**, 784–795.

[14] Sutherland, J. P., 1974. Multiple stable points in natural communities. *Am. Nat.*, **108**, 859–873.

[15] Olson, J. S., 1958. Rates of succession and soil changes on southern Lake Michigan sand dunes. *Bot. Gaz.*, **119**, 125–170.

[16] Clements, F. E., 1916. Plant succession: An analysis of the development of vegetation. *Carnegie Inst. Wash. Pub.*, **242**, 512 pp.

[17] Whittaker, R. H., 1953. A consideration of climax theory: The climax as a population and pattern. *Ecol. Monog.*, **23**, 41–78.

[18] Connell, J. H., and Slatyer, R. O., 1977. Mechanisms of succession in natural communities and their role in community stability and organization. *Am. Nat.*, **111**, 1119–1144; Walker, K. R., and Alberstadt, L. P., 1975. Ecological succession as an aspect of structure in fossil communities. *Paleobiology* **1**, 238–257.

[19] Stone, E. C., and Vasey, R. B., 1968. Preservation of coast redwoods on alluvial flats. *Science*, **159**, 157–161; Florence, R. G., 1965. Decline of old-growth redwood forests in relation to some microbiological processes. *Ecology*, **46**, 52–64.

[20] Beals, E. W., 1965. The remnant cedar forests of Lebanon. *J. Ecol.*, **53**, 679–694.

[21] Watt, A. S., 1947. Pattern and process in the plant community. *Jour. Ecol.*, **35**, 1–22.

[22] Connell, J. H., and Slatyer, R. O., 1977. Mechanisms of succession in natural communities and their role in community stability and organization. *Am. Nat.*, **111**, 1119–1144.

[23] Cushing, E. J., and Wright, H. E., Jr., eds., 1967. *Quaternary Paleoecology*. New Haven, Conn.: Yale University Press; Faegri, K., and Iverson, J., 1964. *Textbook of Pollen Analysis*. New York: Hafner Publishing Co.

[24] Davis, M. B., 1967. Late-glacial climate in northern United States: A comparison of New England and the great lakes region, in Cushing, E. J., and Wright, H. E., Jr., eds. *Quaternary Paleoecology*. New Haven, Conn.: Yale University Press, 11–43.

[25] Wright, H. E., Jr., 1968. The roles of pine and spruce in the forest history of Minnesota and adjacent areas. *Ecology*, **49**, 937–955.

[26] Olson, E. C., 1961. The food chain and the origin of mammals. *Internat. Colloq. on the Evolution of Mammals*; pt. I. Brussels: Kon. Vlaamse Acd. Wetensch. Lett. Sch. Kunsten Belgie, 97–116; Olson, E. C., 1966. Community evolution and the origin of mammals. *Ecology*, **47**, 291–302.

[27] Axelrod, D. I., 1950. Evolution of desert vegetation in North America. *Carnegie Inst. Wash. Pub.*, **590**, 215–360; Axelrod, D. I., 1958. Evolution of the Madro-Tertiary geoflora. *Bot. Rev.*, **24**, 433–509.

[28] Ashby, W. R., 1958. *An Introduction to Cybernetics*. New York: John Wiley & Sons, Inc.

[29] Goodman, D., 1975. The theory of diversity-stability relationships in ecology. *Quart. Rev. Biol.*, **50**, 237–266.

[30] Harrison, G. W., 1979. Stability under environmental stress: Resistance, resilience, persistence, and variability. *Am. Nat.*, 113, 659–669.

[31] Gomez-Pompa, A., Vazquez-Yanes, C., and Guevara, S., 1972. The tropical rain forest: A nonrenewable resource. *Science*, 177, 762–765.

Epilogue

The structured interactions among organisms in natural ecosystems constitute an extraordinarily complex subject. In order to understand a community, you must simultaneously juggle interspecific and intraspecific control mechanisms, abiotic factors, the overall resource base of the ecosystem, and the forces that lead to change in the balances in nature. You must recognize that the community is probably not in equilibrium, but the general principles we derive for ecosystems are commonly based on equilibrium models. There is so much we do not know about the significance of community structure and its components that it is almost frightening.

These ideas come together in the ecosystems that people have manipulated. This has been deliberate, as with agriculture, or coincidental, as with pollution of lakes and streams. The goal in agriculture is to increase the productivity of a single crop species. The agricultural community is unstable, and it has a natural tendency to revert to a more complex community. Pests, crop diseases, weeds, and so on are simply elements in the normal successional sequence of this reversion. Farmers must expend great quantities of time and energy combating it.

The successional imperative is not the only lesson in agriculture. Natural ecosystems are bound together by three classes of signals. The primary system is the concrete phenomena of materials and energy flowing through the system. Certain interactions are based on information, but these are not universal. Finally, there is the flow of genetic material through natural selection. This is what "puts the structure together," as it were, but it is also the slowest and most fraught with delays. These networks of signals are limited. Two are cumbersome, and one is limited in its extent. It is no wonder that ecosystems are sloppy and difficult to comprehend!

Compare the dynamics of the natural ecosystem with agriculture. The farmer uses an option that is not open to natural communities. The hybrid crops that have largely revolutionized modern agriculture are bred as specialists, even in the naturally unstable community of which they are a part. This is possible, however, because farmers can perceive their fields as systems and respond to information signals. They can anticipate future developments and control populations in ways that are inconceivable for systems whose engine of change is natural selection. This does not mean that we can ignore the natural ecosystem; if anything it points out how important it is for us to understand it, since our food supply depends on very close management of a real-world (if artificial) ecosystem.

Further Reading

Allee, W. C., Emerson, A. E., Park, O., Park, T., and Schmidt, K. P., 1949. *Principles of Animal Ecology*. Philadelphia: W. B. Saunders Company.

Begon, M., and Mortimer, M., 1981. *Population Ecology*. Sunderland, Mass.: Sinauer Associates.

Bekoff, M., and Wells, M. C., 1980. The social ecology of coyotes. *Sci. Am.*, **242**(4), 130–148.

Birch, L. C., 1948. The intrinsic rate of natural increase of an insect population. *Jour. Animal Ecol.*, **17**, 15–26.

Blackman, F. F., 1905. Optima and limiting factors. *Annals of Botany*, **19**, 281–295.

Bodde, T., 1981. Ethology: Learning from nature to boost livestock production. *BioScience*, **31**, 629–931.

Bormann, F. H., and Likens, G. E., 1979. Catastrophic disturbance and the steady state in northern hardwood forests. *Am. Sci.*, **67**, 660–669.

Boughey, A. S., 1968. *Ecology of Populations*. New York: Macmillan Publishing Co.

Bretsky, P. W., and Lorenz, D. M., 1970. Adaptive response to environmental stability: A unifying concept in paleoecology. *North Am. Paleont. Conv.* Chicago, 1969, Proc. E, 522–550.

Cameron, G. N., 1972. Analysis of insect trophic diversity in two salt marsh communities. *Ecology*, **53**, 58–73.

Clarke, G. L., 1965. *Elements of Ecology*, rev. printing. New York: John Wiley & Sons, Inc.

Connell, J. H., and Orias, E., 1964. The ecological regulation of species diversity. *Am. Nat.*, **98**, 399–414.

Davis, M. B., 1967. Pollen accumulation rates at Rogers Lake, Connecticut, during late- and postglacial time. *Rev. Paleobotan. Palynol.*, **2**, 219–230.

Davis, M. B., 1969. Palynology and environmental history during the quaternary period. *Am. Sci.*, **57**, 317–332.

Deevey, E. S., Jr., 1947. Life tables for natural populations of animals. *Quart. Rev. Biol.*, **22**, 283–314.

Fager, E. W., 1972. Diversity: A sampling study. *Am. Nat.*, **106**, 293–310.

Fenner, F., 1965. Myxoma virus and *Oryctolagus cuniculus*: Two colonizing species, in Baker, H. G., and Stebbins, G. L. (eds.), *The Genetics of Colonizing Species*. New York: Academic Press, Inc., 485–499.

Finerty, J. P., 1979. Cycles in Canadian lynx. *Am. Nat.*, **114**, 453–455.

Fisher, J., and Peterson, R. T., 1964. *The World of Birds*. Garden City, N.Y.: Doubleday & Company, Inc.

Fox, B. J., 1981. Niche parameters and species richness. *Ecology*, **62**, 1415–1425.

Futuyma, D. J., 1973. Community structure and stability in constant environments. *Am. Nat.*, **107**, 443–446.

Gadgil, M., and Solbrig, O. T., 1972. The concept of r- and K-selection: Evidence from wild flowers and some theoretical considerations. *Am. Nat.*, **106**, 14–31.

Gilpin, M. E., 1975. *Group Selection in Predator-Prey Communities*. Princeton, N.J.: Princeton University Press.

Hairston, N. G., Allan, J. D., Colwell, R. K., Futuyma, D. J., Howell, J., Lubin, M. D., Mathias, J., and Vandermeer, J. H., 1968. The relationship between species diversity and stability: An experimental approach with protozoa and bacteria. *Ecology*, **49**, 1091–1101.

Hairston, N. G., Tinkle, D. W., and Wilbur, H. M., 1970. Natural selection and the parameters of population growth. *Jour. Wildlife Management*, **34**, 681–690.

Hansen, R. M., and Ueckert, D. N., 1970. Dietary similarity of some primary consumers. *Ecology*, **51**, 640–648.

Harper, J. L., 1977. *Population Biology of Plants*. New York: Academic Press.

Holland, R. E., 1969. Seasonal fluctuation of Lake Michigan diatoms. *Limnol. and Oceanog.*, **14**, 423–436.

Hutchinson, G. E., 1959. Homage to Santa Rosalia; or, why are there so many kinds of animals? *Am. Nat.*, **93**, 145–159.

Hutchinson, G. E., 1965. *The Ecological Theatre and the Evolutionary Play*. New Haven, Conn.: Yale University Press.

King, C. E., 1971. Resource specialization and equilibrium population size in patchy environments. *Proc. Nat. Acad. Sci.*, **68**, 2634–2637.

Krebs, C. J., Gaines, M. S., Keller, B. L., Myers, J. H., and Tamarin, R. H., 1973. Population cycles in small rodents. *Science*, **179**, 35–41.

Krebs, J. R., and Davies, N. B., eds., 1978. *Behavioural Ecology: An Evolutionary Approach*. Sunderland, Mass.: Sinauer Associates.

Kushlan, J. A., 1976. Environmental stability and fish community diversity. *Ecology*, **57**, 821–825.

Lawlor, L. R., 1980. Structure and stability in natural and randomly constructed competitive communities. *Am. Nat.*, **116**, 394–408.

Levin, B. R., and Kilmer, W. L., 1974. Interdemic selection and the evolution of altruism: A computer simulation study. *Evolution*, **28**, 527–545.

Levin, S. A., 1972. A mathematical analysis of the genetic feedback mechanism. *Am. Nat.*, **106**, 145–164.

Liebig, J., 1840. Organic chemistry and its application to vegetable physiology and agriculture. Reprinted in abridged form in Kormondy, E. J., 1965. *Readings in Ecology*. Englewood Cliffs, N.J.: Prentice-Hall, Inc., 12–14.

MacArthur, R. H., 1955. Fluctuations of animal populations, and a measure of community stability. *Ecology*, **36**, 533–536.

MacArthur, R. H., 1957. On the relative abundance of bird species. *Proc. Nat. Acad. Sci.*, **43**, 293–295.

MacArthur, R. H., 1972. *Geographical Ecology: Patterns in the Distribution of Species*. New York: Harper & Row, Publishers.

MacArthur, R. H., and Connell, J. H., 1967. *The Biology of Populations*. New York; John Wiley & Sons, Inc.

MacArthur, R. H., and Wilson, E. O., 1967. *The Theory of Island Biogeography*. Princeton, N.J.: Princeton University Press.

Margalef, R., 1963. On certain unifying principles in ecology. *Am. Nat.*, **97**, 357–374.

Margalef, R., 1968. *Perspectives in Ecological Theory*. Chicago: University of Chicago Press.

Mattson, W. J., and Addy, N. D., 1975. Phytophagous insects as regulators of forest primary productivity. *Science*, **190**, 515–522.

May, R. M., 1974. *Stability and Complexity in Model Ecosystems*, 2nd ed. Princeton, N. J.: Princeton University Press.

May, R. M., 1981. *Theoretical Ecology*, 2nd ed. Sunderland, Mass.: Sinauer Associates.

Mayr, E., 1974. Behavior programs and evolutionary strategies. *Am. Sci.*, **62**, 650–659.

McCormick, J., 1968. Succession. *Via*, **1**, 22–35, 131–132.

McCullough, D. R., 1979. *The George Reserve Deer Herd: Population Ecology of a K-Selected Species.* Ann Arbor: University of Michigan Press.

McMillan, C., 1959. Salt tolerance within a *Typha* population. *Am. Jour. Botany*, **46**, 521–526.

McMillan, C., 1960. Ecotypes and community function. *Am. Nat.*, **94**, 246–255.

McNaughton, S. J., 1966. Ecotype function in the *Typha* community type. *Ecol. Monog.*, **36**, 297–325.

McNaughton, S. J., and Wolf, L. L., 1970. Dominance and niche in ecological systems. *Science*, **167**, 131–139.

Murdoch, W. W., 1970. Population regulation and population inertia, *Ecology*, **51**, 497–502.

Murray, B. G., 1979. *Population Dynamics: Alternative Models.* New York: Academic Press.

Nunney, L., 1980. The stability of complex model ecosystems. *Am. Nat.*, **115**, 639–649.

O'Brien, W. J., 1979. The predator-prey interaction of planktivorous fish and zooplankton. *Am. Sci.*, **67**, 572–581.

O'Donald, P., 1972. Natural selection of reproductive rates and breeding times and its effect on sexual selection. *Am. Nat.*, **106**, 368–379.

Odum, E. P., 1969. The strategy of ecosystem development. *Science*, **164**, 262–270.

Odum, E. P., 1971. *Fundamentals of Ecology*, 3rd ed. Philadelphia: W. B. Saunders Company.

Oldfield, J. E., 1972. Selenium deficiency in soils and its effect on animal health, *Geol. Soc. Am. Bull.*, **83**, 173–180.

Paris, O. H., and Pitelka, F. A., 1962. Population characteristics of the terrestrial isopod, *Armadillidium vulgare* in California grassland. *Ecology*, **43**, 229–248.

Phleger, C. F., 1971. Effect of salinity on growth of a salt marsh grass. *Ecology*, **52**, 908–911.

Pianka, E. R., 1970. On r- and K-selection. *Am. Nat.*, **104**, 592–597.

Powell, J. R., and Taylor, C. E., 1979. Genetic variations in ecologically diverse environments. *Am. Sci.*, **67**, 590–596.

Powell, R. A., 1980. Stability in a one-predator three-prey community. *Am. Nat.*, **115**, 567–579.

Reichle, D. E., 1975. Advances in ecosystem analysis. *BioScience*, **25**, 257–264.

Richardson, J. L., 1980. The organismic community: Resilience of an embattled ecological concept. *BioScience*, **30**, 465–471.

Rosenzweig, M. L., 1973. Evolution of the predator isocline. *Evolution*, **27**, 84–94.

Roughgarden, J., 1971. Density-dependent natural selection. *Ecology*, **52**, 453–468.

Silliman, R. P., and Gutsell, J. S., 1958. Experimental exploitation of fish populations. *U.S. Fish and Wildlife Service Fisheries Bull.*, **58**(133), 241–252.

Slobodkin, L. B., 1961. *Growth and Regulation of Animal Populations.* New York: Holt, Rinehart and Winston, Inc.

Smith, O. L., 1980. The influence of environmental gradients on ecosystem stability. *Am. Nat.*, **116**, 1–24.

Southwood, T. R. E., 1977. Entomology and mankind. *Am. Sci.*, **65**, 30–39.

Suter, G. W., II, 1981. Ecosystem theory and NEPA assessment. *Bull. Ecol. Soc. America*, **62**, 186–192.

Tanner, J. T., 1966. Effects of population density on growth rates of animal populations. *Ecology*, **47**, 733–745.

United Nations, 1970. *Demographic Yearbook*, 21st ed., 1969. New York: Statistical Office of the United Nations, Dept. of Economic and Social Affairs.

Vitousek, P. M., and Reiners, W. A., 1975. Ecosystem succession and nutrient retention: A hypothesis. *BioScience*, **25**, 376–381.

West, D. C., Shugart, H. H., and Botkin, D., 1981. *Forest Succession: Concepts and Applications*. New York: Springer-Verlag.

Whittaker, R. H., 1970. The population structure of vegetation. In Tüxen, R., ed. *Gesellschaftsmorphologie (Strukturforschung)*. The Hague: Verlag Dr. W. Junk N. V., 39–62.

Woodwell, G. M., and Smith, H. H., 1969, eds. *Diversity and Stability in Ecological Systems*. Brookhaven Nat. Laboratory Symp., **22**.

Natural
Ecosystems

IV

THE VARIETY in the biosphere is so great as to be virtually incredible. There are many different kinds of ecosystems, and they all show significant differences, as well as significant similarities in the way they function. The previous parts of this book have presented the principles of ecosystem operation. This part will describe a range of ecosystems and integrate the principles with the actual geographic conditions to which they respond.

Aquatic ecosystems are those based on water, whereas terrestrial ecosystems are based on land. They are as different from each other as any two kinds of ecosystems could be, yet they have some important points in common. Water is the medium of nutrient transport in both types of systems. This is obvious in aquatic ecosystems: Water is the medium for the ecosystem. But it is no less true in terrestrial ecosystems. The difference is that, here, the water that carries the nutrients is the interstitial water in the soil, and its flow and the way that nutrients are held in bio-available form are both quite different from those of an aquatic ecosystem.

Solid materials take a different form in the two ecosystems. Soils are the most important type in terrestrial ecosystems, they are not found in aquatic ecosystems. Soil may wash into streams or lakes, but it loses the characteristics of soil and merges with other solid materials washed in from elsewhere in the bottom sediments. Nevertheless, the bottom sediments play a role in the mineral nutrition of the aquatic ecosystem almost as great as does the soil in terrestrial ecosystems. It is impossible in either instance to understand the workings of the ecosystem without understanding the interrelationships between water and solid materials and between these nonliving components and life.

The atmosphere does not impinge as directly onto aquatic ecosystems as it does onto terrestrial ecosystems. But the air-water interface is a critical surface, since life could not persist without oxygen and nitrogen. This is almost as true for terrestrial ecosystems as for lakes and streams. Despite the fact that most animals and the trunks and leaves of plants are surrounded by the atmosphere, the roots are not. The level of aeration of the roots often determines the difference between survival and death for plants.

As we describe the different kinds of aquatic or terrestrial ecosystems, you should keep in mind that the basic metabolism of organisms in the two environments is not very different. Their adaptations to their environments may be quite different, however, as are the responses of their ecosystems to different sorts of signals.

Aquatic Ecosystems

7

LIQUID WATER COVERS about three quarters of the earth's surface, almost all as oceans. Virtually all surface waters contain life in one form or another; hence, aquatic ecosystems would be important for their sheer volume, if for nothing else. Even now, photosynthesis in tiny single-celled marine plants is the main force maintaining the earth's atmospheric oxygen. In addition, aquatic ecosystems are simpler in many ways than terrestrial systems, since the omnipresent factor that sets the tone for all aquatic ecosystems, regardless of their biotic complexity, is water. It is the medium within which all aspects of the ecosystem coexist, both living and nonliving. It is the source of all nutrients for aquatic life, including the gaseous nutrients such as oxygen and carbon dioxide. It is the medium by which organic and inorganic wastes and sediments are distributed throughout the ecosystem. The amount of light energy reaching the community depends on how much light is absorbed by water; the heat properties of water control the circulation patterns within the ecosystem and are a major influence on the structure of the aquatic community that exists in any one place.

There are many kinds of aquatic ecosystems, but the most meaningful breakdown is based on the salinity, or amount of material dissolved in the water. Most of the earth's surface water (99.99%) is in the oceans, which contain about 35 parts per thousand (‰) dissolved salts. Of the remainder, most is fresh water with a salt content under $\frac{1}{5}$ ‰ found in either lakes and ponds (still water) or in rivers and streams (running water). The most common transition zone between a river and the sea is the *estuary*, which has a dissolved solid content intermediate between those of fresh and marine waters. Finally, some surface waters are *hypersaline*, with a dissolved salt content substantially higher than that of the open ocean. These occur commonly in arid areas, where rivers flow into lakes that have no outlet, so that the salt in the inflowing rivers is concentrated as the water evaporates. Hypersaline water can also be found in nearshore lagoons and oceanic embayments with restricted connection to the open ocean (e.g., the Red Sea). Here, marine water is trapped behind physical barriers and salts are concentrated through evaporation.

Water and Aquatic Ecosystems

Aquatic ecosystems are simpler than terrestrial ecosystems. This is not because the interactions among their components is any different, but rather because water provides such a powerful unifying force to all aquatic ecosys-

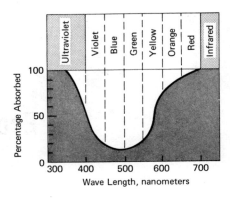

Figure 7.1
Absorption of light by 10 meters of pure water as a function of wavelength. [Reprinted, with permission, from P. K. Weyl, *Oceanography: An Introduction to the Marine Environment*. Copyright © 1970 by John Wiley & Sons, Inc., New York.]

tems. Water has the same general relationship to the other environmental factors in the ocean as it does in a lake, a stream, or an estuary. The properties of water control the productivity, life style, and complexity of the aquatic environment in ways for which there is no analogue in terrestrial communities.

Light Penetration in Water

Water absorbs light, transforming radiant energy into heat. Both ends of the visible light spectrum, especially the red end, are selectively absorbed, however (Figure 7.1), so that the appearance of daylight changes with increasing depth from white to bluish to dull blue-green. As in the atmosphere, absorption by water is not the only means by which light levels are reduced: Light may be diffused or absorbed by suspended particles such as sediments, detritus, animals, and plants. The more particles in the water, the more diffusion and absorption.

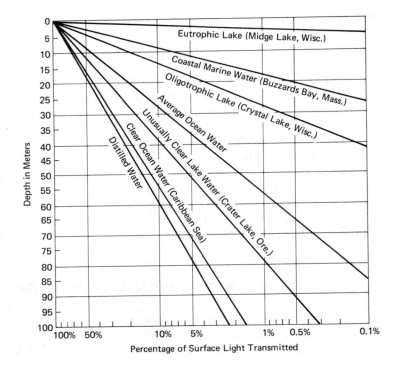

Figure 7.2
Rate of absorption of light in different bodies of water. The curves are based on the yellow-green (500–600 nm) portion of the spectrum. [Adapted, with permission, from G. L. Clarke, *Elements of Ecology*, rev. ed. Copyright © 1954 by John Wiley & Sons, Inc., New York.]

Light heats the environment as it is absorbed, and it powers aquatic photosynthesis. The rate of absorption of light by water is much faster than that by air. Figure 7.2 indicates the amount of light transmission in ocean water under various conditions of turbidity. Even in the cleanest of water, the percentage of light absorbed by 6 to 8 m. of water is equal to that of the entire atmosphere. In ecosystems containing a large complement of suspended particles, virtually all of the light is absorbed within a few centimeters.

Because light is absorbed so much faster by water than by air, it is often a key limiting factor in an aquatic ecosystem (unlike terrestrial systems). A plant must fix more energy than it utilizes in its respiration in order to survive. That is, the ratio between a plant population's gross primary production and respiration must be greater than 1. There is a level in the water at which the mean daily light flux allows a rate of production equal to that of respiration. This is the *light compensation level*. Above this level, autotrophs can produce sufficient food for themselves and the animals associated with them. Below it, at least some energy must be derived from organisms living and feeding in the lighted zone and sinking below it, either as part of their normal behavior or after death. The light compensation level is not a well-defined surface. It varies seasonally, tending to be higher in high latitudes and in winter, since the total radiation flux is lower than in low latitudes or in summer. Because it is the depth at which the most efficient plant species is limited by light, it depends on the makeup of the community, especially the relative efficiency of the plants in utilizing light energy. It is also higher in waters containing a high concentration of light-scattering materials. In general, however, it is the depth penetrated by about 1 to 5% of the light that reaches the surface.

Solubility of Gases in Water

Most gases dissolve readily in water, most notably those that are essential for life. The concentration of any gas in water generally varies between zero and *saturation*. This is the amount of gas that is dissolved when the atmosphere and the water are in equilibrium with one another. It is easy to get air and water in equilibrium in the laboratory; one simply puts some water into a flask and shakes it up. But except for waterfalls and very turbulent streams, the water in natural ecosystems is seldom in equilibrium with the atmosphere. Concentration of a gas may be in *deficit*, if it is being utilized in the ecosystem faster than it is being dissolved across the air-water interface, or it may be *supersaturated*, if it is being produced in the ecosystem faster than it is being released from solution across the interface. The concentration of important gases may vary widely in any ecosystem, both horizontally and vertically. The rate at which gases cross the air-water interface depends on the extent of the water surface and on its turbulence.

The saturation level of any gas in water depends on several variables, most notably temperature, salinity, the concentration of the gas in the atmosphere, and its relative solubility in water. In general, the higher the temperature or dissolved salt concentration of an ecosystem, the lower the saturation concentration of a gas in it (Figure 7.3). The greater the concentration of a gas in the atmosphere, the greater its concentration in the water will tend to be, depending on its relative solubility in water. Thus, carbon dioxide is about two hundred times as soluble as oxygen in water, and oxygen is about

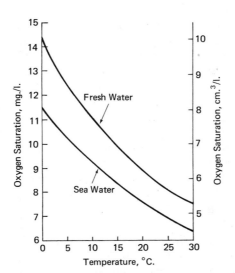

Figure 7.3
Curves for oxygen saturation with respect to temperature in fresh water and in sea water.

twice as soluble as nitrogen. But because the atmosphere contains 3.7 times as much nitrogen as oxygen, the saturation level for oxygen is lower than that for nitrogen. For carbon dioxide, it is lower still (Table 7.1). The makeup of essential gases in aquatic ecosystems is very different from those of terrestrial ecosystems. When compared to nitrogen, both oxygen and carbon dioxide are relatively more abundant in water at saturation. Carbon dioxide, in fact, is as abundant in water as it is in air in absolute terms. The absolute oxygen concentration in aquatic ecosystems, however, is so much lower than in the atmosphere that it is often limiting, especially when oxygen concentration is significantly below saturation.

OXYGEN. From the viewpoint of living organisms, no gas is more important than oxygen. It constitutes 21% of the atmosphere by volume: That is, a liter of air contains 210 cc of oxygen gas. The saturation level of dissolved oxygen in water is about 10 mg/l (7 cc/l) at 14°C, and varies with temperature as shown in Figure 7.3. Thus oxygen is about 30 times more abundant in the air than it is at saturation in water. This difference is even greater under conditions of oxygen deficit. Because all heterotrophs use oxygen for their respiration, it tends to be removed continuously from solution. The only means by which it can enter the water are through solution at the air-water interface or through photosynthesis by aquatic plants. Oxygen is transferred to deeper water by either diffusion or circulation of surface water.

The rate of oxygen solution at the surface depends mainly on the turbulence of the water and the presence of waves, spray, and foam. A typical lake can absorb about 100 g/m²/day (± 60%) at standard atmospheric pressure. As small as this sounds, it is sufficient to saturate a water column 10 m in depth, and is about 6,200 times greater than the average amount of oxygen produced by photosynthesis in open water. Diffusion downward from the sur-

Gas	Atmospheric Concentration	Saturation in Water
Oxygen	210 cc./l. (21%)	7 cc./l. (32.9%)
Nitrogen	780 cc./l. (78%)	14 cc./l. (65.7%)
Carbon Dioxide	0.3 cc./l. (0.03%)	0.3 cc./l. (1.4%)

Table 7.1
Comparison Between Equilibrium Concentrations of Various Important Gases in the Atmosphere and in Water (in cc./l. and Percentage of Total Gas Concentration)

Figure 7.4
Proportions of various forms of carbon dioxide over the range of pH commonly found in ecosystems. [Reprinted, with permission, from G. E. Hutchinson, *A Treatise on Limnology; Vol. 1: Geography, Physics, and Chemistry.* Copyright ⓒ 1957 by John Wiley & Sons, Inc., New York.]

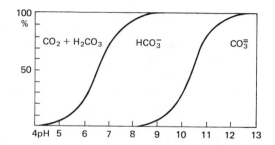

face is an extremely slow process, and the oxygen concentration at the bottoms of all bodies of standing water would probably be nil if diffusion were the only means to get oxygen to deep levels. Circulation mixes oxygenated surface water with deeper water. The oxygen concentration can become uniform from the top to the bottom of a body of water when mixing is complete. However, as we shall see, mixing is seldom complete in natural bodies of standing water, and oxygen consumption in the waters below the light compensation level often produces a substantial oxygen deficit in deeper waters.

NITROGEN. Nitrogen is significantly less soluble in water than oxygen. But it still accounts for about 65% of the dissolved gases at equilibrium, because of its abundance in the atmosphere. It is fairly inert chemically and does not react with water, although nitrogen-fixing bacteria, fungi, and blue-green algae, can use it to make ammonia or nitrate, and other bacteria can denitrify fixed nitrogen under conditions of very low oxygen concentration.

CARBON DIOXIDE. The chemistry of carbon dioxide in water is somewhat more complex than that of either oxygen or nitrogen, because carbon dioxide and water react to form carbonic acid and carbonate ions (page 54). The chemical form of dissolved carbon dioxide is related to pH, as indicated in Figure 7.4.

DISSOLVED SOLIDS IN WATER. The solid materials dissolved in water are exceedingly variable, not only in distribution but also in origin and importance. Of all surface waters, only the ocean approaches uniformity of dissolved solids; its salinity varies between 30 and 37‰, with virtually all of it falling between 34 and 36‰. The dissolved salts include the ions shown in Table 7.2. There is no equivalent uniformity for fresh waters: There is no "typical" lake or river. Some rivers with very low dissolved salt concentration have a salinity as low as 40 parts per million (ppm), as in the Savannah River of Georgia. The Colorado River at the Grand Canyon has a dissolved

Table 7.2
Composition of Sea Water (35‰): Major Ionic Components

Reprinted, with permission, from G. Dietrich, *General Oceanography: An Introduction.* Copyright 1963 by John Wiley & Sons, Inc., New York.

Chloride (Cl⁻)	19.345‰
Sulfate (SO₄⁻)	2.701
Bicarbonate (HCO₃⁻)	0.145
Bromide (Br⁻)	0.066
Sodium (Na⁺)	10.752
Magnesium (Mg⁺⁺)	1.295
Calcium (Ca⁺⁺)	0.416
Potassium (K⁺)	0.390

solid concentration of 800 ppm. Still others, such as Oklahoma's Cimarron, are so hypersaline that salt crystals precipitate into their beds.

There are several sources for nutrients and other materials in fresh waters. Dust particles washed out by rain is often a significant one. Salt crystals thrown up from the sea in spray are carried by the wind, and raindrops form around them, dissolving the salt crystals. In industrial areas, gaseous pollutants such as sulfur dioxide, nitrogen oxides, and various acids, are adsorbed onto solid particles of flyash. The pollutant dissolves in the rainwater when raindrops form around the flyash. Much of the dissolved solids in freshwater ecosystems comes from surface runoff. These solids may include organic compounds containing nutrient material, salt excreted by plants and animals, or weathering products from the chemical breakdown of rocks. They may enter a river as soluble materials or as solid compounds that are broken down to soluble form by the detritus food chain in the stream. Finally, groundwater reaching the surface contains materials dissolved from rocks and soil layers. The most common materials contributed by seepage and springs are silica (SiO_2), carbonates, and readily soluble salts gained by solution of subsurface rock. This source may be relatively unimportant, as in areas where the permeability of rocks is low, or it may be very important in areas where aquifers are extensive and contain soluble salts (this is, for instance, the source of salt in salt flats of the Cimarron and several other rivers of the arid Southwest).

The essential elements are all found as dissolved salts. Only in this form can they be absorbed by aquatic organisms. As in any ecosystem, some are present in relatively large quantities, and others are quite rare. How different nutrients limit productivity differs from place to place. The most consistently suboptimal nutrient in most freshwater ecosystems appears to be phosphorus, followed by nitrogen, sulfur, and calcium. Nitrogen appears to limit productivity in marine ecosystems, at least those nearshore environments where experimentation is possible (see page 50 for discussion of limiting factors).

The concentration of an essential element in the water is closely related to the activity of living organisms. Nutrient concentration reaches a maximum during the winter, when production and the rate of withdrawal of nutrients from solution by living organisms is at its minimum. It falls sharply as the rate of withdrawal increases and algal growth spurts in the spring (Figure 7.5). It is then released back into the water as organisms die and are broken down by detritus organisms. The potential for tremendous fluctuation of nutrient levels in the ecosystem is clear. Natural selection in many species of algae has led to the development of mechanisms by which limiting nutrients (notably phosphate) are stored in excess of the short-term needs of the algae. It is thus available in the population when dissolved nutrient concentrations bottom out. Absorption of excess nutrient, or *luxury uptake*, is a very effective internal means by which algal populations can even out some of the variations in their abiotic environment.

Phosphorus is the most important nutrient limiting overall aquatic productivity in freshwater ecosystems. But it is not the only one. Nitrogen is also a common limiting nutrient, and other elements may limit particular groups of organisms. David Tilman's experiments on diatoms[1] are a classic example of the interaction of silica and phosphorus as limiting factors (see also page 85). When the dissolved silica concentration falls too low to main-

Figure 7.5
The relationship between nutrient concentration and productivity: phosphate content drops sharply as diatoms increase in the early spring. Following the spring diatom increase, the phosphate content remains low, as other organisms remove phosphate from solution. Data are from Loch Striven, Scotland. [Reprinted, with permission, from W. D. Russell-Hunter, *Aquatic Productivity*. Copyright © 1970 by The Macmillan Company, New York.]

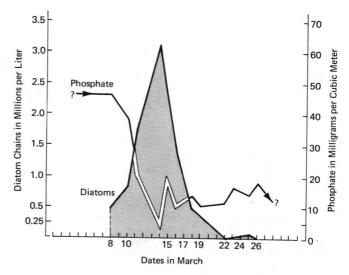

tain diatom growth, the diatoms may be replaced by much less desirable algae such as blue-green algae.[2] One element may limit productivity under some circumstances and another under others. Limiting nutrients are used by living organisms in various organic molecules, and the ratio of nitrogen to phosphorus in living tissues is about 7:1 by weight. Nitrogen is typically about 20 times more abundant than phosphorus in fresh water. In addition, several species of aquatic bacteria, fungi, yeasts, and blue-green algae can fix dissolved nitrogen.

The critical role of limiting nutrients in aquatic ecosystems is underscored by their presence in domestic sewage. Nutrient-bearing discharges of domestic sewage into watercourses has, in essence, fertilized surface waters, so that the productivity of moderately polluted aquatic ecosystems may be very high. As we shall see (page 195), the consequences of this increase in productivity are not always desirable, at least from the viewpoint of the human observer. Of course, sewage effluents beyond a certain level constitute a stress on the ecosystem, and productivity falls off again.

A final illustration of the role of limiting nutrients in aquatic ecosystems is the very high productivity in marine zones of upwelling, where nutrient-rich cold currents from the ocean depths come to the surface to displace nutrient-poor surface waters (see page 71).

Aquatic Habitats

Living organisms are very unevenly distributed in aquatic ecosystems. There are several basic life styles, each of which is related to the organism's location within the ecosystem (Figure 7.6).

Some unattached organisms live at the air-water interface. These include floating plants, as well as many types of animals. Some animals spend their lives on top of the interface (e.g., water striders), while others (e.g., diving beetles and back-swimmers) spend most of their time beneath the interface and obtain much of their food from within the water. Collectively, this life form is called *neuston*. It is an important life-style in freshwater ecosystems, but it is quite unimportant in the oceans.

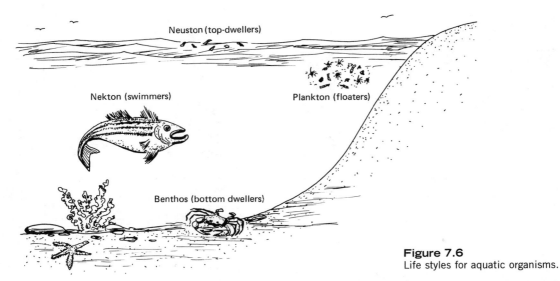

Figure 7.6
Life styles for aquatic organisms.

Plankton include the small plants and animals found in all aquatic ecosystems except for fast-moving rivers. Their powers of self-locomotion are limited, so that their distribution is controlled largely by currents in their ecosystems. Most can move a bit, however, either to control their vertical distribution or to seize prey. Most planktonic plants, or *phytoplankton*, as well as all planktonic animals, or *zooplankton*, are capable of at least some motion. Certain zooplankton are exceedingly active and move relatively great distances considering their small size, but they are so small that their range is still controlled largely by currents.

Some animals in almost all aquatic ecosystems are swimmers, or *nekton*. In order to overcome currents, these animals are relatively large and powerful. They range in size from the swimming insects of quiet water, which may be only about 2 mm long, to the largest animal that has ever lived on earth, the blue whale.

The *benthos* includes the organisms living at the bottom of the water mass. They are particularly interesting in the variety of their adaptations to the environment. The bottom is a more heterogeneous habitat than either the open water or the surface, and this variety is reflected in the organisms. Some organisms live *on* the bottom, and others live *in* the bottom sediments. Virtually every aquatic ecosystem contains a well-developed benthos. The adaptations of organisms in benthic communities reflect the composition of the bottom, its stability or tendency to shift, and its depth.

Role of the Watershed in Aquatic Ecosystems

The watershed of a lake or stream is the area it drains. Water running over the surface ultimately enters the lake or stream in question, just as water reaching the surface from a subsurface reservoir. Except for lakes with very long retention times (and, of course, the oceans), the flux of nutrients and energy in aquatic ecosystems is determined largely by the materials washed in from their watersheds. The surrounding areas are a critical factor in determining the characteristics of lakes and streams. In many cases, the responses of an ecosystem to a particular stimulus are due more to what goes on in the watershed than to the behavior of the biota in the lake or stream. Most important, perhaps, if we wish to alter an aquatic ecosystem (especially one that has become polluted), the most effective way to do this is to control the materials that enter the ecosystem from outside.

Moving-Water Ecosystems

Moving-water ecosystems include rivers, streams, and related environments. They are remarkably variable, ranging in size from the Amazon River, whose mean rate of flow is about 93,000 m^3/sec to the trickles of a small spring. They vary from raging torrents and waterfalls to rivers whose flow is so smooth as to be almost unnoticeable. The characteristics of any river vary greatly over its length, as it changes from a tiny brook to its full size. All aspects of the ecosystem change, from abiotic factors to community. In general, the community is the most useful indicator of change, but its usefulness stems from the way it responds to change in abiotic factors and integrates the myriad changes in this realm.[3]

The most distinctive features of moving-water ecosystems are those related to their motion. Most obvious are the rate of flow and the stream velocity. The rate of flow, or *discharge*, refers to the volume of water passing a given observation point during a specific unit of time; it is measured in units such as m^3/sec, ft^3/sec, or acre-feet/sec[4] Discharge tends to increase steadily going downstream, as tributaries join with the main river. The *velocity of flow* is the speed at which the water moves, and is measured in m/sec, ft/sec, or mi/hr. Velocity is more variable than discharge.

One of the most significant factors in moving-water ecosystems is the turbulence, or the irregularity of motion of the water particles. Flow can be perfectly even, or *laminar*: Water particles move parallel to one another, and mixing is minimal. This contrasts with *turbulent flow*, where movement of water particles is highly irregular (Figure 7.7). Turbulent water has great erosive power and *shear forces* at the water-sediment interface, and the amount of oxygen incorporated into the water is very high. Streams with laminar or near-laminar flow have lower erosive power and shear forces at the stream bottom, and relatively less oxygen is incorporated into the water.[5]

These are the factors that mold the environment. The rate at which oxygen is incorporated into a stream determines the level of the oxygen deficit. This is probably the most significant limiting factor for the stream community. The shear forces are those forces that remove particles from the stream bed, be they sediment, plants, or animals. This means, among other things,

Figure 7.7
Diagram showing the differences between turbulent (a) and laminar (b) flow of water in a stream.

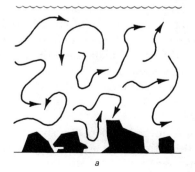
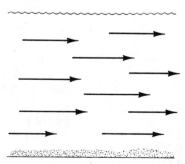

a   b

that the shear force is what must be overcome by any benthic organism if it is to maintain its habitat. This is a major factor for a large portion of the moving-water community.

The erosive power of the stream, which is related to the shear forces, has many roles in the river. First, it determines the rate at which the landscape surrounding the stream is sculptured. Within the stream, it controls the types and diversity of habitats by determining which sediment particles are carried by the stream and which are left behind in the stream bed. If a lot of material is being carried, the turbidity of the water is high, and light penetration is reduced. Conversely, if erosion by the stream is such that the water is clear, light may penetrate to the bottom. All the abiotic variables of the moving water ecosystem form gradients from some minimum value to a maximum. Every environment within each river is slightly different, but most streams share certain general trends. Enough streams show these trends that we can construct a "textbook model," which can summarize the changes that take place in real rivers.

Our river begins in a highland as a small, fast-moving, clear, cold, turbulent brook. Its flow gradient is steep, and it actively erodes its bank. The stream bed is composed of large rock fragments, worn smooth and round by the action of the water. Going downstream, the river is joined by tributaries. In most areas, except where mountains butt up against the coast, the river passes from the highlands into a rolling hill land such as the high plains of western North America or the piedmont of the East. The gradient is lower, although the stream flow is still rapid and turbulent. The pebbles composing the stream bed are much smaller, and there are occasional areas where the bottom consists of sand that has settled out of suspension. The water is more turbid than it was in the highland, because there are more very fine sediment particles in suspension, but it is still very clear.

As the river flows on to the sea, its gradient decreases even more as it crosses the coastal plain. The bed is sandy or silty here, and deposition of sediment is more common than erosion. The discharge is higher because the river has been joined by many tributaries, but the channel is likely to be much deeper. The water is quite turbid, because it is full of very small sediment particles drawn from the entire watershed of the river, as well as organic material and living organisms. Turbulence tends to be low, and flow may be laminar. Finally, the river reaches the ocean, either directly or through an estuarine transition zone.

The river shows a continuum of environments from one end to the other, and none is "typical" of the entire river. It is most convenient to pick two specific ecosystems: the rapidly flowing portion and the slowly flowing portion. These terms are deliberately relative, and they cannot be defined precisely. They vary in their usage from river to river (e.g., a slowly flowing river by Rocky Mountain or Alpine standards would probably be considered a raging torrent on the coastal plain), and their main purpose is to provide a framework for viewing some of the differences between organisms and communities inhabiting the rivers.

Rapidly Flowing Water

Rapidly flowing water is the part of a river where flow is both rapid and turbulent. Everything that is not attached or weighted is swept away by the current, including organisms and sediment particles alike. The substrate tends to be rock or gravel, and the fragments are rounded and smoothed by the water. The habitat itself is extremely diverse. The dominant physical factor in the environment is the shear force exerted by the moving water. But shear force and rate of water movement tend to be quite different on top of a rock fragment, between rock fragments, and beneath rock fragments. Different species can exploit these differences in microhabitat.

The animals that live on the tops of rocks need an efficient mechanism for staying in one place to avoid being swept away by the current. In fact, a great many individuals are swept away even in species with the most effective adaptations for staying put. The organisms in this microhabitat are universally flattened. Some, like the freshwater limpet, are virtually flat. They hold themselves in place with a very large powerful foot that extends over almost the entire area of the shell and offer little resistance to the current. The water penny is the larva of the riffle beetle; not only is it almost flat, but hooked claws on each of its legs enable it to hold onto the substrate more firmly. Freshwater sponges actually cement themselves to the surface of the rocks. Caddis flies build "houses" out of sediment or wood fragments, which are then cemented firmly to the rocks.

Among the most common organisms on the exposed rock surfaces are sessile algae. There are often so many of them that they make the rocks feel slippery. But few plants can survive in the rapid-water environment, since the shear force at the rock-water interface is so great that only small, well attached forms can survive. The energy base for the metabolism of this ecosystem is organic detritus washed into the river from its watershed. This is far more important in the overall picture of the stream than in-stream photosynthesis. In consequence, most primary consumers are detritus-eaters.

Several different kinds of animals live in the spaces between the rocks. Many of them, like mayfly and stonefly naiads, are flattened and have behavioral adaptations to hold them in place. They cling instinctively to any hard surface (such as a rock, another insect larva, or even, in times of great desperation, a limnologist's finger). They also orient themselves to face into the current and move upstream. The result of this combination in many of these naiads is that the current presses the insects tightly against the rock, increasing the friction between them and their substrate. Other insect larvae, such as the hellgrammite, are large and covered with spines. Their size makes it somewhat harder for the current to sweep them away, and the spines help in holding them in place between the stones.

Many species, such as flatworms, annelid worms, other insect larvae, clams, and some species of snails, live beneath the rocks. The current is weaker here, and animals are less likely to be carried away. The basic adaptations to this habitat are similar to the others in the fast-water ecosystem, but they need not be quite as clearly drawn. The overlap in microhabitat between rapid-water animals is considerable, and many species can be found in several different microhabitats. But the diversity within the fast-water benthos is high, resulting largely from the physical differentiation of the habitat.

If there are areas where the current is slow enough, swimming organisms such as fish can be found. The fish of fast-water ecosystems tend to be specialized cold-water species such as salmon or trout, which are highly prized by people. In many fast-water ecosystems, however, the current is simply

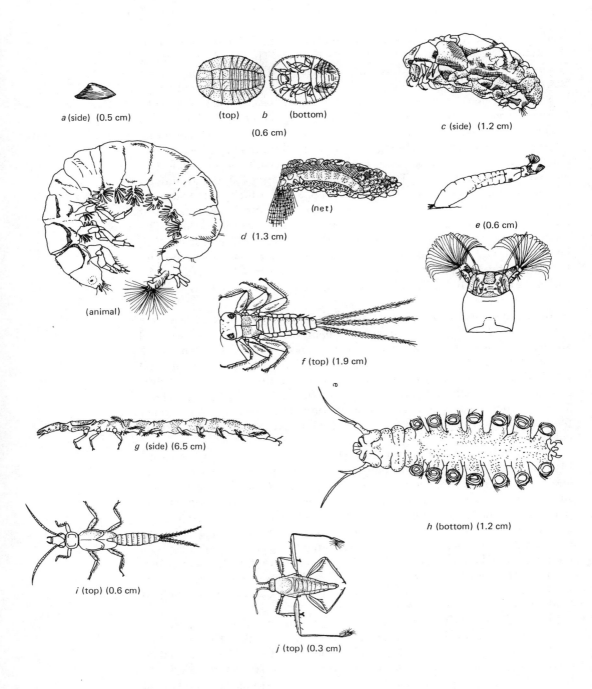

a (side) (0.5 cm)

(top)   b   (bottom)
(0.6 cm)

c (side) (1.2 cm)

(net)

d (1.3 cm)

e (0.6 cm)

(animal)

f (top) (1.9 cm)

g (side) (6.5 cm)

h (bottom) (1.2 cm)

i (top) (0.6 cm)

j (top) (0.3 cm)

Figure 7.8
Organisms found in rapidly flowing water: (*a*) *Ferrissia,* a small freshwater limpet; (*b*) *Psephenus,* the water penny; (*c*) *Glossosoma,* a caddis fly larva with a case made of small stones; (*d*) *Hydropsyche,* a caddis fly larva commonly found in rapidly moving water without a case, but which spins a net to catch food flowing downstream; (*e*) *Simulium,* a blackfly larva; with large head fans which catch food flowing downstream; (*f*) *Stenonema,* a mayfly naiad with a highly flattened body (note the three tails and the row of gills along the sides of the body; this is typical of mayflies); (*g*) *Corydalus,* the hellgrammite, the larva of a dobson fly (note the spines along the sides of the body); (*h*) *Deuterophlebia,* a fly larva with a double row of suckers extending most of the length of its body; (*i*) *Chloroperla,* a stonefly naiad (note the two tails; this is typical of stoneflies); (*j*) *Rhagovelia,* a water strider which lives in unusually turbulent water (note the bristle on the last segment of the middle leg).

too fast for any swimming organisms. In such cases, the benthos may be rich and diverse, but it may also be the entire community.

Fast-water ecosystems have much in common with cold, deep lakes. Water temperatures and productivity tend to be low, and diversity is high. In fast-water ecosystems, the main control of productivity is the current, which limits the amount and type of autotroph production that can take place.

Slowly Flowing Water

A slowly flowing stream is a very different sort of environment from the one just described. Because the flow is both slower and more likely to be laminar, the erosive power of the stream is greatly reduced, and smaller sediment particles are deposited on the bottom instead of being carried away by the stream. The variegation of the mud-bottom habitat is much lower than that of rocky fast streams. Some bottom-dwelling organisms may clamber over the bottom, such as isopods (sowbugs), snails, and mayfly and damselfly naiads. Others burrow into the sediment, including tubeworms, naiads of the burrowing mayflies, and several other insect larvae, as well as clams, nematodes, snails, and rotifers. Abundant swimming organisms include not only fish (which tend to be different species from those of fast-water areas), but also large crustacea (e.g., freshwater shrimp) and many types of insects. Several insects alternate between crawling and swimming. Finally, several insects spend most of their time at the surface of the stream. These include water striders, water boatmen, back-swimmers, and predaceous diving beetles. Zooplankton are common, including a rich assemblage of protozoa and smaller crustacea such as cladocera (water fleas) and copepods.

Plant life is abundant in a slow-water ecosystem. It includes rooted vascular plants such as pond weeds and grasses, firmly attached aquatic mosses, and large multicellular filamentous algae. Very small floating plants such as duckweed may cover much of the surface of the slowly moving stream, especially in its slowest backwaters. Motile algae such as diatoms and flagellates may abound in the open water. The primary productivity of a slow-water ecosystem is higher than that of the rapid-water system, and the community is relatively less dependent on food materials from outside. Even so, a substantial portion of the community's energy is washed in as detritus. In addition, the normal detritus food chain consisting of organisms such as bacteria and fungi is much better developed in this community than it is upstream, as the partially decomposed organic debris that constitutes the main food source for these organisms accumulates in the mud bottom. In some slow-moving streams, in fact, the bottom muds contain more organic material than mineral fragments.

If the main controlling factor in fast-water streams is the current, the dissolved oxygen concentration is the main limiting factor in slow-water streams. The high level of animal activity, coupled with an active detritus food chain, can withdraw a large amount of oxygen from the water. In addition, the low level of turbulence means that less oxygen is incorporated into the water at the surface. The dissolved oxygen concentration of a slowly moving stream can be substantially less than saturation, and the community must be much more tolerant of low oxygen conditions as a result. For example, the trout and salmon found in fast water require high oxygen levels,

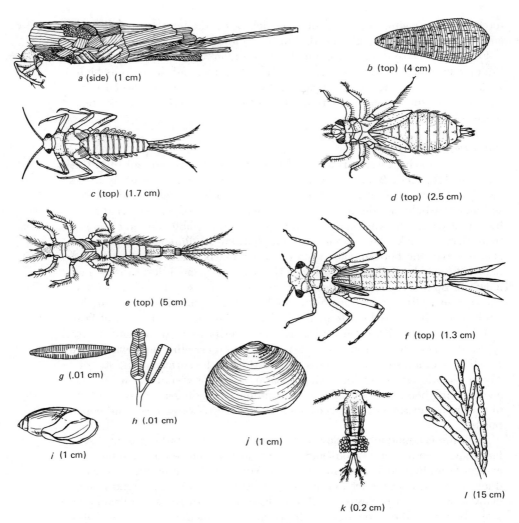

Figure 7.9
Organisms found in slowly flowing water. Many of these are also found in ponds, and many orgaisms whose prime habitat is ponds are also found in slowly flowing water: (a) *Anabolia*, a caddis fly larva with a case made of twigs; (b) *Glossiphonia*, a leech which lives under stones; (c) *Baetis* a mayfly naiad; (d) *Paragomphus*, a burrowing dragonfly naiad; (e) *Hexagenia*, a burrowing mayfly naiad (the adults of this genus are known as the Canadian Soldier and were at one time so common as to be a pest near the Great Lakes); (f) *Ishneura*, a damselfly naiad; (g) *Navicula*, a planktonic diatom; (h) *Gomphonema*, an attached diatom; (i) *Physa*, a pulmonate snail; (j) *Sphaerium*, the fingernail clam; (k) *Cyclops*, a copepod (note the egg sacs); (l) *Cladophora*, an attached green alga which often forms long streamers on rocky surfaces.

while the most common fish of slow water are often tolerant species such as carp and catfish.

Lakes as Environments

Lakes are bodies of standing fresh water. They may be shallow or deep, large or small. In any case, the features that give them their most distinctive characteristics stem from the fact that the water is standing. A typical lake has one or more inlets and an outlet (some very tiny lakes and ponds may not have an inlet stream, and salt lakes may have no outlets, but they are exceptional). A typical moderate-sized or large lake has several different communities in the various subregions. Except for shallow ponds, lakes tend to be *thermally stratified*. As with moving-water ecosystems, we can make certain generalizations about lakes, but each lake is different, and there are exceptions to every rule, especially when a lake is either very small or very large.

Thermal Stratification in Lakes

The physical properties of a substance seldom exert more direct control on the characteristics of an ecosystem than in the case of thermal stratification of lakes. The maximum density of pure water is reached at 3.94°C; all fresh water at any other temperature found on the surface of the earth is lighter. Therefore, all fresh water of any temperature will float on water whose temperature is 3.94°C unless there is some force sufficient to mix the layers and bring about a uniform temperature.

Consider a deep lake whose temperature is fairly uniform from top to bottom—about 4°C. This is a normal condition in temperate-zone lakes during part of the spring. As the day lengthens, solar radiation is absorbed by the water. Most of the absorption takes place in the uppermost layers of the lake. If all the absorbed radiation is transformed into heat (a good assumption), and if the water is perfectly still (a bad assumption), we could expect the surface of the lake to be quite warm, and the temperature to be decreasing gradually with depth to 4°C (Figure 7.10). The level below which no light penetrated (i.e., the level below the compensation level) could be heated only by diffusion, which is negligible. Thus the compensation level for photosynthesis and the lower level of heat absorption would be essentially the same.

This model does not hold, since the surface waters of a lake are not perfectly still. There are several forces acting on them, of which the most important is the wind. This mixes the surface water to a depth of 8 to 20 m., but it does not reach down to the deeper water levels. The result is a warm top layer heated by the sun and homogenized by the wind and other currents, underlain by a cool deeper layer that is not heated by the sun and is too deep to be circulated directly by the wind. The upper layer is called the *epilim-*

Figure 7.10
Theoretical distribution of temperature in a lake during the summer if there were no mixing of water by the wind.

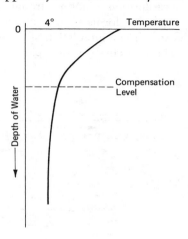

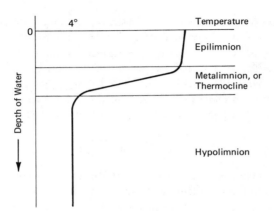

Figure 7.11
Typical pattern of temperature distribution in a deep lake in the temperate zone during the summer.

nion, and the lower layer is called the *hypolimnion*. The transition between the two is the *metalimnion* or *thermocline* (Figure 7.11).

There is no necessary relationship between the compensation level and any of these layers. In a typical lake, photosynthesis can proceed in the entire epilimnion, but it is excluded from hypolimnion. However, the compensation level may be below the thermocline in unusually clear lakes, and it is likely to be above it in turbid lakes. Stratification is maintained on strictly thermal bases, since warm water is lighter than the colder water. Once a thermocline has formed, the two layers are virtually separate from one another. The current patterns are also different, although they may be related. Virtually no water passes from the epilimnion to the hypolimnion or vice versa (Figure 7.12).

As fall progresses, heat is lost from the epilimnion to the atmosphere faster than it is absorbed, and the temperature drops. Finally, the epilimnion reaches a point where it is essentially the same temperature as the hypolimnion. The thermocline no longer exists to prevent circulation between the upper and lower layers; hence the entire lake can mix from top to bottom. This period of lakewide circulation is called the *overturn*.

As the regional temperature continues to drop, water at the top of the lake continues to lose heat and becomes less dense than the lower layers. A thermocline tends to reform with a cold epilimnion (< 4°C) overlying a "warm" (4°C) hypolimnion. In typical temperate-zone lakes, the winter stratification is not as strongly developed as the summer stratification, because ice forms if the temperature at the surface drops 4° or more, and wind-driven circulation of the epilimnion ceases. Thus the temperature distribution curve for a typical winter stratification resembles the theoretical

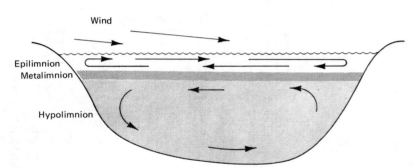

Figure 7.12
Diagrammatic illustration of the patterns of wind-induced circulation in the hypolimnion and epilimnion of a thermally stratified lake. Circulation in the hypolimnion is much less pronounced than that in the epilimnion.

Figure 7.13
Seasonal progression of thermal stratification patterns in a typical first-class lake through a year's cycle in the temperate zone.

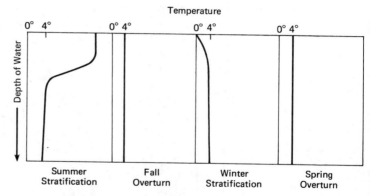

Temperature

| Summer Stratification | Fall Overturn | Winter Stratification | Spring Overturn |

curve for temperature distribution in the absence of agitation more than the curve for summer stratification.

With the onset of spring, the lake again absorbs heat from the sun, and its temperature rises until it is uniform about 4°C from top to bottom, the lake turns over, and the cycle begins again. Figure 7.13 summarizes the thermal cycle for a typical lake of the temperate zone.

Not all lakes undergo two overturns per year. A cold stratification cannot develop in areas where the regional temperature never falls below 4°C, and a warm stratification can never develop when the regional temperature never rises above 4°C. Overturn happens only once in either of these cases. In the former, which is common in tropical areas, water turns over only during the coldest season; in the latter, which is common in arctic areas, water overturns only during the warmest season (Figure 7.14).

The distribution of heat in the various strata is related both to lake size and regional climate. Lakes can be classified into three categories, based on their heat relations. *First-class lakes* store virtually all their heat uptake in the epilimnion. These are deep lakes in which the hypolimnion remains essentially uniform at 4°C all year round, and the overturn takes place for a short period when the epilimnion is also at 4°C. *Second-class lakes*, store a significant amount of heat in the hypolimnion as well as in the epilimnion. Thus the summer temperature of the hypolimnion is significantly greater than 4°C, although it is still cold enough to maintain thermal stratification. These

Figure 7.14
Seasonal progression of thermal stratification patterns in warm and cold lakes with only one overturn through a year's cycle.

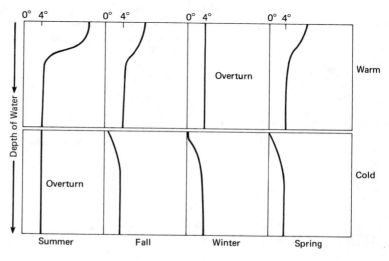

| Summer | Fall | Winter | Spring |

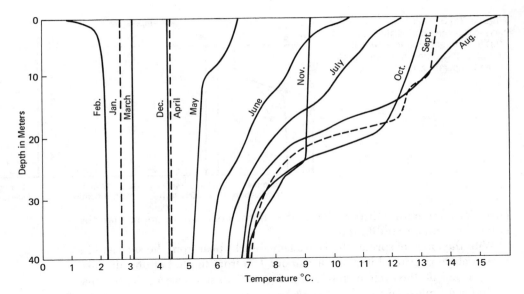

Figure 7.15
Annual temperature progression in a second-class lake—Convict Lake, California: monthly variation of thermal stratification; [Redrawn from N. Reimers and D. Combs, "Method of Evaluating Temperatures in Lakes with Description of Thermal Characteristics of Convict Lake, California," *U.S. Fish and Wildlife Service Fisheries Bulletin,* **56** (105), 535–553.]

lakes are substantially shallower than first-class lakes, and more heat is supplied by the environment than can be taken up in the epilimnion. Because the hypolimnion changes temperature as well as the epilimnion, the entire lake is of uniform temperature for a greater length of time than in a first-class lake (Figure 7.15). The period of overturn in a second-class lake thus lasts longer than in a first-class lake. *Third-class lakes,* on the other hand, tend to be so shallow that winds can mix them to the bottom, and they do not stratify.

Life Zones in Lakes

Different parts of a lake have different characteristics, and hence support different communities. Differences stem from several factors, including substrate, light, thermal stratification, and geographic position within the lake. To be sure, some species are found in more than one life zone. But the distinctions between the life zones are as well defined as any in nature. A

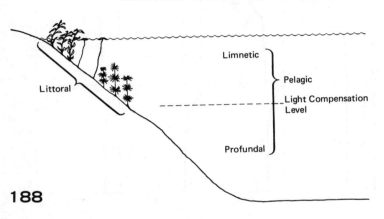

Figure 7.16
Diagram of the main life zones of a lake.

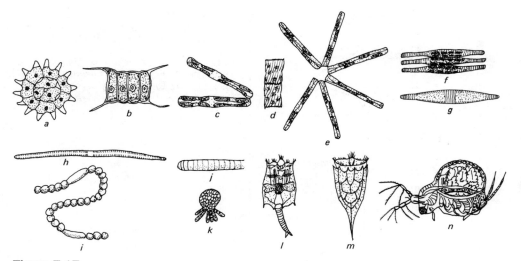

Figure 7.17
Plankton of the limnetic zone of lakes: (a) *Pediastrum*, a colonial green alga; (b) *Scenedesmus*, a colonial green alga; (c) *Tabellaria*, a colonial diatom; (d) *Melosira*, a colonial diatom; (e) *Asterionella*, a colonial diatom; (f) *Fragillaria*, a colonial diatom; (g) *Nitzschia*, a diatom; (h) *Synedra*, a diatom; (i) *Anabaena*, a blue-green alga; (j) *Oscillatoria*, a blue-green alga; (k) *Difflugia*, a protozoan related to *Amoeba*, but with a shell composed of tiny pieces of sediment; (l) *Brachionus*, a rotifer; (m) *Keratella*, a rotifer; (n) *Moina*, a cladoceran, or water flea. All are extremely small microorganisms.

meaningful and useful framework for describing the life zones in aquatic ecosystems is shown in Figure 7.16.

The open-water, or *pelagic* zone is the area where the bottom is too deep to be inhabited by rooted plants. It can be subdivided by the light compensation level into a lighted zone in which autotrophs can produce food through photosynthesis, and a dark zone in which photosynthesis is not sufficient to support a community. In the upper (*limnetic*) zone, the community structure is normal: Autotrophs are abundant and form the base of the food web. Single-celled planktonic algae found there are the main source of primary production for the lake as a whole. The diversity of life is impressive. The plankton alone includes dinoflagellates, green and blue-green algae, several sorts of crustacea, protozoa, and rotifers. There are also some fish. The lower (*profundal*) zone cannot support photosynthesis, and the main

Figure 7.18
Benthic invertebrates found in the profundal zone of lakes: (a) *Mysis*, a mysid crustacean; (b) *Tubifex*, the tubeworm of sludgeworm; (c) *Valvata*; (d) *Bithynia*; (e) *Musculium*, the Pea-shell clam; (f) *Chironomus*, the bloodworm, a midge larva.

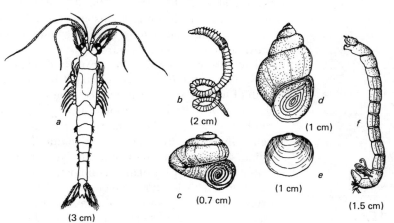

(a) (3 cm)
(b) (2 cm)
(c) (0.7 cm)
(d) (1 cm)
(e) (1 cm)
(f) (1.5 cm)

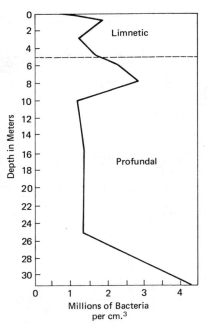

Figure 7.19
Bacterial population distribution at the height of summer stratification in Lake Glubokoye, U.S.S.R. [Redrawn, with permission, from G. E. Hutchinson, *A Treatise on Limnology. Vol. 1: Geography, Physics, and Chemistry.* Copyright © 1957 by John Wiley & Sons, Inc., New York.]

source of energy is detritus that sinks out of the limnetic zone. All the organisms are chemosynthetic autotrophs or heterotrophs, either as detritus feeders or as carnivores. Most of our largest lake fishes inhabit the dark waters of the hypolimnion for most of their lives. In addition to detritus from the limnetic zone, materials washed in by rivers can settle out and serve as food for profundal animals.

The lake bottom in young lakes may be of the original rock; in older lakes, it is covered with sediment to form a uniform substrate of mud or sand. The profundal benthos of a typical lake is not a particularly diverse community, but is an important one. It includes several species of insect larvae, including those of small mosquitolike midges and certain burrowing mayflies, clams, snails, and the ubiquitous tubeworms. In addition, there is a well-developed series of decomposer bacteria, extending from the top to the bottom of the profundal zone (Figure 7.19). Many species of plankton inhabit the profundal zone for part of the day and then migrate into the limnetic zone. These vertical migrations are controlled by the amount of light in the ecosystem and may lead to an upward movement in the morning and downward at night, or vice versa.

The area shoreward of the lower limit of rooted vegetation is more complex. This is the *littoral* zone, and it has the greatest diversity in the lake, both of habitat and of community. The portion that is always under water is the zone of rooted vegetation. In the deepest part of this zone, the plants tend to be completely submerged. The "zone of submergent vegetation" is generally fairly wide, and it may include the entire bottom of shallow lakes, if light can penetrate to the bottom. Nearer to shore is the zone populated by water lilies and other plants with floating leaves. Closest to shore is the zone of emergent vegetation, which is characterized by plants with leaves that extend above the surface of the water, such as cattail (*Typha*), bulrush (*Scirpus*), and arrowhead (*Sagittaria*). The large numbers of plants allow primary productivity to be high, but they also act as an efficient trap for organic sediments and nutrient materials washed in from land.

In addition to rooted vegetation, the littoral zone also contains virtually all the kinds of phytoplankton found in the limnetic zone, as well as some other algae that are not found in open water. Many littoral algae are *epiphytic*: They are attached to rooted plants and do not normally float freely. These include numerous species of diatoms, green algae, and blue-green algae.

The animals of the littoral are as diverse as the plants. Many are *epizoic*, and live their lives attached to the stems or leaves of rooted plants. These include protozoa, rotifers, insect larvae, hydras, and bryozoa. Other animals (e.g., snails, flatworms, and various types of insect larvae and nymphs) are not attached to plants but spend their lives moving over them. Some of these animals are herbivorous, some are carnivorous, and still others feed on the particulate detritus that swirls through the vegetation in the eddy currents set up by the wind and the waves. Littoral zooplankton include many species found in the limnetic zone, as well as some that are not. The latter include larger forms that rest on the rooted vegetation while they are not actively swimming.

The littoral is rich in small swimming animals, especially insects. These include back-swimmers, diving beetles, water boatmen, and similar forms, many of which are also found in streams. Many minnows are exclusively littoral, as are relatives of the sunfish, and even some large carnivores like the northern pike. Many other species of fish may pass freely from the littoral to the pelagic, and others spend their larval period in the littoral and their adulthood in the pelagic. Other amphibious vertebrates such as frogs, turtles, and snakes, and mammals such as muskrats and beavers are found almost exclusively in the littoral, when they are not on land. The animal benthos is also extremely diverse. Many groups are represented, both living within the sediment and moving over it. Most are detritus feeders that feed on the extensive quantity of organic particles that get trapped around the rooted vegetation, but some are carnivores.

Distribution of Dissolved Oxygen and Nutrients in Lakes

The productivity of a lake depends on the oxygen and available nutrients dissolved in the water. The epilimnion of a lake is typically well oxygenated, because of photosynthesis and solution of oxygen across the air-water interface. Dissolved nutrient concentrations are highly variable, since the nutrients in the epilimnion are constantly cycled back and forth between the water column and living tissue, especially those in low supply. Actual production in the epilimnion is controlled by the amount of available nutrients and oxygen in the water column. Nutrient concentrations are controlled primarily by the patterns of cycling within the epilimnion, the rate at which nutrient-rich nutrients are lost to the hypolimnion, the rate at which soluble nutrients return from the hypolimnion during overturn, and any other mechanism by which nutrients are recycled within the ecosystem.

Nutrients cycle rapidly between the community and the water column (page 63), and they are also recycled from sediments into open water in the littoral zone. All green plants can produce photosynthetic oxygen, but only rooted plants can draw nutrients such as phosphorus out of the sediments through their roots and release them into the water.[6] As a result, the dynamics of productivity differ considerably between a lake whose littoral zone is

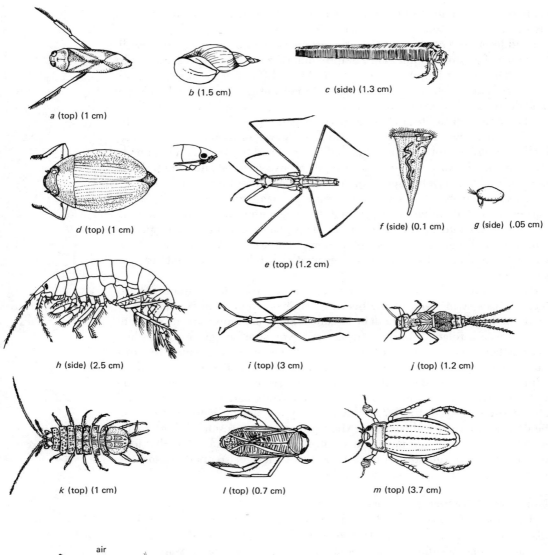

a (top) (1 cm)

b (1.5 cm)

c (side) (1.3 cm)

d (top) (1 cm)

e (top) (1.2 cm)

f (side) (0.1 cm)

g (side) (.05 cm)

h (side) (2.5 cm)

i (top) (3 cm)

j (top) (1.2 cm)

k (top) (1 cm)

l (top) (0.7 cm)

m (top) (3.7 cm)

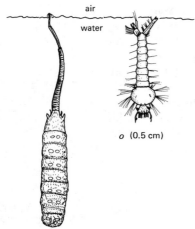

air

water

o (0.5 cm)

n (1.3 cm)

large in relation to its pelagic zone and one whose pelagic zone is large in relation to the littoral. The larger the littoral, the more sediment is trapped by rooted plants, and the more nutrient recycling from sediments to water column exists.

In contrast, only heterotrophic metabolism proceeds in the hypolimnion. This means that hypolimnion production is determined by the amount of detritus raining into it. Not only is detritus the only source of energy, it also (and perhaps more importantly) controls the concentration of oxygen and nutrients in the hypolimnion. If productivity in the epilimnion is low, the amount of detritus sinking into the hypolimnion is also low. Hence decomposer metabolism is low, a minimum of oxygen is withdrawn from hypolimnion water to support detritus breakdown, and a minimum of soluble nutrient is released into the hypolimnion from the detritus food chain. If production is somewhat higher, the detritus rain is still fairly low, but a buildup of organic detritus on the bottom leads to the withdrawal of a significant amount of oxygen from the lowermost levels. If production is high, the detritus rain is substantial, and a large amount of oxygen is withdrawn from the water throughout the hypolimnion.

The demand for oxygen by detritus organisms is greatest at the bottom-sediment surface where detritus accumulates, and the dissolved oxygen concentration at the bottom of a lake may commonly reach zero, so that anaerobic decomposition becomes dominant. Because the decomposition of detritus releases soluble nutrients, there is an inverse relationship between oxygen concentration in the hypolimnion and the concentration of vital nutrients.

Productivity Classification of Lakes

It is paradoxical that the greater the productivity of a lake, the lower the oxygen concentration in its hypolimnion. But this is the case for all stratified lakes, and it has ramifications on aquatic community structure that might not be expected from simple data on productivity alone. An *oligotrophic* lake is one whose productivity is low (from the Greek, "poorly nourished"). Figure 7.21a shows the distribution of the temperature and oxygen in an extremely oligotrophic lake, which is virtually sterile biologically. The water is essentially saturated with oxygen from top to bottom, although a small amount of oxygen is withdrawn from the hypolimnion. Figure 7.21b shows another oligotrophic lake, although with sufficient production that the amount of bacterial respiration in the sediment is significant. The hypolimnion is still

Figure 7.20 (opposite)
Invertebrates found in the littoral zone of lakes: (a) *Notonecta,* the backswimmer, a predaceous bug; (b) *Lymnaea,* a pulmonate snail; (c) *Brachycentrus,* a caddis fly larva whose case is constructed from thin woody scraps (the case is square in cross-section); (d) *Dineutus,* the whirligig beetle (this insect has split eyes; one section is normally above water, the other below the water line); (e) *Gerris,* the common water strider; (f) *Stentor,* a large ciliate protozoon; (g) *Cypris,* the ostracod; (h) *Gammarus,* an amphipod, or fresh-water "shrimp"; (i) *Ranatra,* a water scorpion, a predaceous bug; (j) *Caenis,* a mayfly naiad (note the large flaps formed by one pair of gills which cover much of the dorsal surface of the abdomen; this structure is related to the mechanism by which the animal obtains oxygen in its silty habitat); (k) *Asellus,* the sowbug, a fresh-water isopod; (l) *Corixa,* the water-boatman, an herbivorous bug; (m) *Dytiscus,* the predaceous diving beetle; (n) *Eristalis,* the rat-tailed maggot (the long "tail" is actually a breathing tube which extends above the surface of the water. This animal is especially common in water with a substantial oxygen deficit); (o) *Culex* larva, the common mosquito.

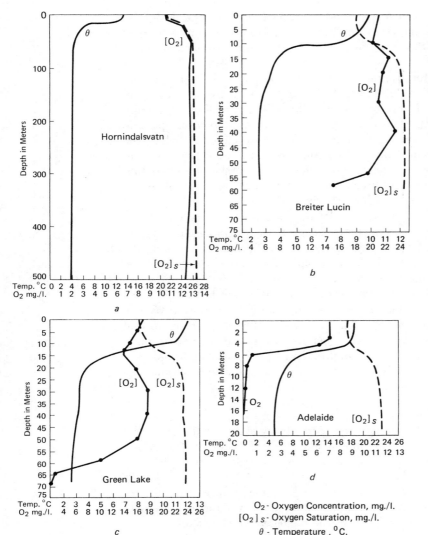

Figure 7.21
Temperature and oxygen distribution curves in several lakes: (a) Lake Hornindalsvatn, Norway, a biologically sterile mountain lake; (b) Lake Breiter Lucin, north Germany, an oligotrophic lake with some oxygen depletion at the bottom; (c) Green Lake, Wisconsin, a eutrophic lake; (d) Adelaide Lake, Wisconsin, a eutrophic virtually no oxygen in the hypolimnion. [Redrawn, with permission, from G. E. Hutchinson, *A Treatise on Limnology; Vol. 1: Geography, Physics, and Chemistry.* Copyright © 1957 by John Wiley & Sons, Inc., New York.]

O_2 - Oxygen Concentration, mg./l.
$[O_2]_s$ - Oxygen Saturation, mg./l.
θ - Temperature , $^\circ$C.

well oxygenated, but there is significant oxygen withdrawal near the bottom. A *eutrophic* lake is one in which production is high (from the Greek, "well nourished"). Figures 7.21c and d show the distributions of temperature and oxygen for two typical eutrophic lakes. There is substantial oxygen depletion from the top of the hypolimnion to the bottom, with depletion most extreme in the bottom layers.

Oxygen is thoroughly mixed and is essentially at saturation in most lakes during overturn. This pattern remains throughout the year in oligotrophic lakes. But the hypolimnion oxygen concentration in eutrophic lakes drops progressively after stratification is established, until very little oxygen remains below the thermocline (Figure 7.22). Concurrent with the depletion of oxygen in the hypolimnion, the breakdown of organic material leads to the release of nutrients such as phosphorus. Figure 7.23 compares the distribution of phosphate in oligotrophic Crystal Lake, Wisconsin, and eutrophic Linsley Pond, Connecticut.

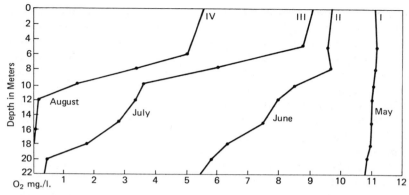

Figure 7.22
Removal of oxygen from the hypolimnion in Lake Mendota, Wisconsin, following spring overturn. [Reprinted, with permission, from G. E. Hutchinson, *A Treatise on Limnology; Vol. 1: Geography, Physics, and Chemistry.* Copyright © 1957 by John Wiley & Sons, Inc., New York.]

Eutrophication and the Life Cycle of Lakes

Lakes are ephemeral features when viewed in geologic time. They are born when a quirk of history determines that part of a river's drainage will be wider and deeper than the rest. As sediments build up in the lake, the bottom rises until the lake disappears altogether, leaving an organic soil within the river's flood plain as its only record.

A geologically young lake typically has a low concentration of dissolved nutrients. Gross productivity is limited as a result, and it is a typical oligotrophic lake. The animals that inhabit its lower reaches have little tolerance of low oxygen conditions; these include active fishes, burrowing mayflies, bloodless midge larvae, and mollusks. The phytoplankton are diverse, consisting mainly of diatoms and other golden-brown algae. As time progresses, nutrient materials accumulate in the lake, either as substances dissolved in river water or as solids in the sediments carried by the rivers. Some nutrients are also washed out of the atmosphere by rain.[7] For some shallow lakes the atmospheric nutrient contribution alone is sufficient to cause nutrient enrichment problems. Productivity increases as the nutrient level in the lake rises, until it reaches a maximum level typical of the eutrophic condition.

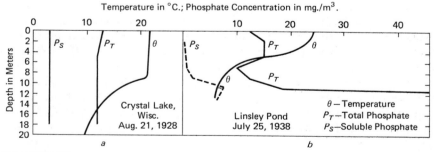

Figure 7.23
Vertical distribution of phosphate in lakes: in oligotrophic Crystal Lake, Wisconsin (*a*), and in eutrophic Linsley Pond, Connecticut (*b*). [Reprinted, with permission, from G. E. Hutchinson, *A Treatise on Limnology; Vol. 1: Geography, Physics, and Chemistry.* Copyright © 1957 by John Wiley & Sons, Inc., New York.]

Eutrophic communities are very different from those of an oligotrophic lake. The active fishes intolerant of low oxygen levels are gone, replaced by so-called coarse fishes such as perch and carp. The only animals that can survive in the benthos are those that can obtain oxygen from virtually deoxygenated water, such as hemoglobin-containing tubeworms (tube-dwelling relatives of the earthworm) and bloodworms (midge larvae). Some species of bloodworm can respire normally if the oxygen concentration is as low as 1% of saturation (about 0.1 ppm). They can survive in lakes where relatives that can obtain adequate oxygen from water at 2% of saturation cannot.

The phytoplankton are largely green and blue-green algae, and they are subject to periodic algal blooms, in which these algae expand from a relatively low concentration to such abundance that the surface of the water looks as if it is covered with the skim layer of cheap green oil paint. Under normal circumstances, the high level of production in the epilimnion provides a lot of detritus to the hypolimnion, leading to oxygen depletion. An algal bloom represents a great increase in production for the lake as a whole. When the algae die, however, there is a tremendous increase in detritus raining into the hypolimnion. This leads to an even greater loss of oxygen in the hypolimnion. Algal blooms are commonly associated with polluted bodies of water (with reason), but they were characteristic of eutrophic lakes long before modern sewage-producing people came on the scene. Oneida Lake in New York, for instance, was so prone to blooms of blue-green algae that even the Indians of the area used to call it "Stinking Green" in the eighteenth century.

The development of a lake from oligotrophic to eutrophic is a much more complex process than a simple increase in productivity. This is the process of *eutrophication*. It also involves drastic changes in the roles of various environmental factors, and it commonly results in an ecosystem whose biota is very different from the original. It is generally believed that eutrophication is a normal feature of lakes which necessarily takes place as they fill with sediment. Natural eutrophication has been observed many times for small lakes. Eutrophication has also been observed in large lakes. However, the nutrient addition that drives the eutrophication of large lakes tends to be of human origin, and the process is *cultural eutrophication*. This has become a serious problem around the world.[8] The nutrients are added along with domestic sewage, and all sectors of the economy contribute to it in some degree. It, in turn, affects recreation, fisheries, water supply, and public health. Algal blooms reach nuisance proportions in lakes in which they were once unknown. Fish populations are excluded from lakes or even driven into extinction by the reduction of dissolved oxygen in the hypolimnion. Some bacteria commonly found in eutrophic lakes cause public health problems. Eutrophication brings an increase in total available nutrients, especially phosphorus, through time. The available phosphorus is very different from the accumulation of all the phosphorus that has entered the lake, however. It depends on the interactions among organisms, air, water, bottom sediments, and the surrounding watershed (Figure 7.24). Some of these, such as the transfers across the air-water interface and the contribution from the watershed, do not change much in a natural ecosystem. Others, such as the dynamics of nutrient transfer across the sediment-water interface and the phosphorus balance within the community, show progressive change as the lake ages.

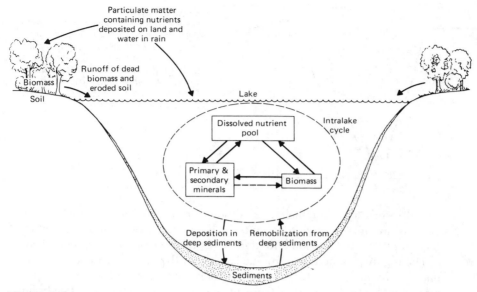

Figure 7.24
Processes determining flow of nutrients such as phosphorus into and within aquatic ecosystems.

Productivity in the oligotrophic lake is limited by the low level of available nutrients. As the nutrient pool rises, so does production. But why should the available nutrient concentration in an oligotrophic lake rise? It is not altogether obvious that it should, since bottom sediments perform a scavenging function to remove phosphorus from biological availability. Several alternative hypotheses suggest why the nutrient content of a lake increases. For example, there may be a decrease in the sediments' scavenging ability. Phosphorus is more soluble and less easily adsorbed under the reduced-oxygen conditions that accompany eutrophication. In addition, the siltation and consequent increases in shallow water area may allow an expansion of rooted plants that are capable of pumping phosphorus out of the sediments, even when it is buried sufficiently deeply that it would otherwise be unavailable. Either mechanism would allow the flux of phosphorus through the biota to rise with the aging of a lake even if the only physical change were sedimentation to the lake bottom and even if there were no change in the net influx of phosphorus from the atmosphere or the watershed.

Other potential controlling mechanisms lie outside the lake. The typical watershed may supply phosphorus to the lake faster than it can be removed in sediments and outflow. This kind of situation is not uncommon, and it places control of the rate of eutrophication for such a lake largely on characteristics of the watershed. In fact, paleolimnological studies have demonstrated that natural changes in nutrient delivery patterns in streams can lead to acceleration and reversals in the eutrophication rate.[9]

The changes that accompany eutrophication are not completely reversible. Any changes in the input of a limiting nutrient affects production. They will also affect the composition of the biota and the scavenging behavior of the sediments. Returning a key variable such as nutrient influx from the watershed to its original level imposes a change on an altered ecosystem in response to the new conditions. There is no reason to assume that the lake

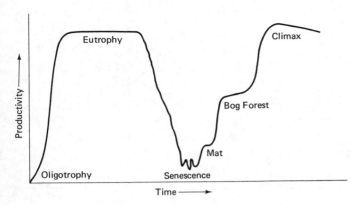

Figure 7.25
Generalized development of the productivity of a lake through the lake cycle, from oligotrophy to eutrophy to dystrophy, then eventually to a stable climax as the regional climax forest. [Reprinted from R. L. Lindeman, "The Trophic-Dynamic Aspect of Ecology," *Ecology*, **23** (1942), 399–418.]

would return to its original state, although it might. The behavior of nutrients, communities, and sediments in lakes is complex, and the status of these entities at any point in time is a function of their overall history.

The highly productive eutrophic lake is not the end of its development. It may be the longest lasting stage, but sooner or later, the sediment carried into the lake builds up to a point where the lake changes drastically once again. Old lakes typically do one of two things. They may simply fill up, perhaps through a bog succession (page 153), or they may proceed to a *dystrophic* (from the Greek, "ill-fed") condition before filling in completely. In the latter instance the lake's oxygen deficit is very high, as it contains enough organic material to deoxygenate the entire lake. Anaerobic breakdown of detritus produces the highly acid "brown water" typical of many swamps. Because of the very harsh conditions arising from low oxygen concentration and low pH, the community—insofar as it exists at all—is very simple, consisting of many individuals of a few species. It is an exceedingly unstable community, and production is low.

Eventually, the dystrophic lake fills with sediment and terrestrial succession begins. Production is higher in this phase, reaching a stable maximum as a climax terrestrial community. The changes in productivity through the various stages of the ontogeny of a lake are shown in Figure 7.25. Not all lakes follow this precise path, and the time scales involved may be quite variable. A small lake may pass from oligotrophy through eutrophy through dystrophy to climax forest in a few thousand years. The huge Lake Baikal in the USSR is still oligotrophic after nearly 100 million years.

Estuaries

Estuaries are the most common ecosystems of transition between rivers and the sea. They are semienclosed coastal waterways with free connection to the open sea, and within which seawater is diluted with fresh water from the rivers. Not all rivers open into estuaries; some simply discharge their runoff into the ocean, where it may either keep its integrity for some distance offshore or mix with the seawater relatively quickly.

The estuary is not simply a transition zone between fresh and marine water. It is also an extremely significant ecosystem type in its own right. Regardless of their size or location, estuaries around the world have two things in common. First, they are semienclosed, so that marine currents do

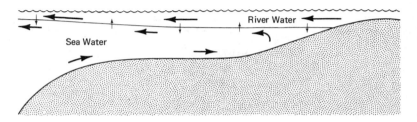

Figure 7.26
Theoretical flow of fresh water over sea water in an area where a river enters the sea without tidal or other effects to mix fresh water and sea water.

not control the estuarine environment; second, the opening to the sea is sufficiently free that the estuary is subject to tides. Because of these two characteristics, the estuary is truly a hybrid between the river, all of whose discharge passes through it, and the ocean, which contributes both seawater and tidal currents.

Fresh water has a lower density than seawater (± 1.00, as opposed to 1.03). This difference is much greater than the density differences that result in the thermal stratification of lakes (it is, in fact, roughly equivalent to the density difference between freshwater masses at 4° and 80°C.). Were there no tides in an estuary to mix fresh and salt water, the lighter fresh water would simply flow over the heavier seawater and dissipate in the ocean (Figure 7.26). However, the tidal action acts as a plunger to flush the estuary and mix the fresh and salt water (Figure 7.27). This has several results. The water level in the estuary fluctuates with the tides, unlike that of a river. The habitat that is covered at high tide and uncovered at low tide is a prominent one in estuaries, and it has no analogue in any purely freshwater ecosystem. Furthermore, the salinity is exceedingly variable and may change by a factor of ten over the course of a day at any location.

At low tide, most of the water passing through the estuary is fresh river water, and the salinity is correspondingly low. At high tide, most of the water may be of marine origin and the salinity correspondingly high (Figure 7.28). Flow may be somewhat stratified, or mixing may be incomplete, and the salinity at any time may vary greatly from one place to another, even over short distances. The intensity of the current, and hence of the degree of mixing, depends on the intensity of the tides and of the river's flow rate. The physical environment of the estuary is exceedingly variable. No organism that cannot tolerate great salinity changes, currents, and turbidity could hope to survive in it.

Nevertheless, an estuary's high productivity is exceeded among aquatic communities only by coral reefs and oceanic zones of upwelling. Although an estuary is a variable environment, its turbulence leads to a fairly high dissolved oxygen concentration. In addition, the tidal action concentrates the nutrient and energy materials that wash in from upstream, or, in some cases, that enter from the nutrient-rich bottom waters of the sea. Particulate nutrient material enters the estuary at its upper end, is carried seaward by the

Figure 7.27
Diagram showing the effects of tides on mixing fresh river water with sea water.

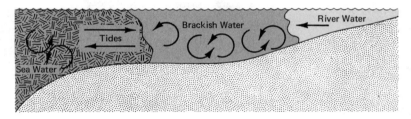

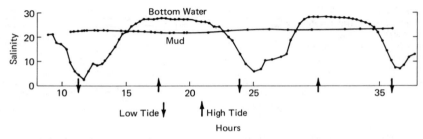

Figure 7.28
Fluctuation of salinity in an estuary as a function of time and tide. Data are from the Pocasset River estuary, Massachusetts. [Reprinted, with permission, from P. C. Mangelsdorf, "Salinity Measurements in Estuaries," in G. H. Lauff (ed.), *Estuaries,* A.A.A.S. Pub. No. 83 (1967), 71–79. Copyright © 1967 by the American Association for the Advancement of Science, Washington.]

falling tide, and then is brought back through the estuary by the rising tide, and so on for several cycles. It takes much longer for a nutrient particle to traverse an estuary than to pass through a similar length of even the most slowly flowing river. Thus, the estuary acts as a nutrient trap, with an average nutrient level substantially higher than either the river or the sea that it connects. In the same way, the concentration of energy-rich organic materials is very high.

The result of this concentration of nutrients and fixed carbon is a very high level of production within the detritus food chain. Nutrient-bearing materials are broken down bacterially at a very high rate and recycled into soluble form. Plants adapted to the rugged current conditions and salinity found in the estuary can maintain a high level of productivity. The amount of nutrient-rich organic detritus also allows a very high level of productivity for detritus-eating animals.

The estuarine community is a mixture of three components: the marine, the freshwater, and the brackish-water. As shown in Figure 7.29, the diversity of both freshwater and marine components drops in the estuary, and the brackish-water component reaches a maximum, but overall estuarine diver-

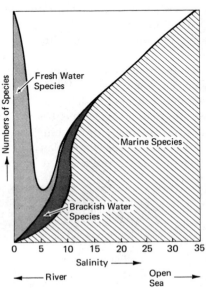

Figure 7.29
Numbers and types of species inhabiting an estuary. Data are from the Bay of Kiel, Germany. [Reprinted after Remane from K. O. Emery and R. E. Stevenson, "Estuaries and Lagoons," *Geological Society of America Memoir 67,* Part 1, 535–586.]

sity is still lower than that of the river or marine community. This is not surprising, considering the tremendous variation in the estuary's physical environment. But it means that the great productivity of estuaries is built on a narrow base.

The plants of the estuary are of four basic sorts: phytoplankton, marginal and marsh vegetation, mud-flat algae, and epiphytic plants growing on the marginal marsh vegetation. Because of the turbid water found in estuaries, phytoplankton are normally uncommon, as the compensation level tends to be quite high. Their populations are exceedingly variable, however, and great blooms of certain algae are well known. Most estuarine algae are of marine origin. The most obvious estuarine plants are the marginal and marsh vegetation. These include mangroves and marsh grasses, as well as some submerged filamentous colonial green algae. Few animals feed on these plants directly, but a very large amount is consumed as detritus. The mud flats which are uncovered at low tide may be sites of intense photosynthesis by diatoms and filamentous blue-green algae. The brown color of a mud flat may be due as much to numbers of diatoms as to the presence of organic material in the mud. The epiphytes include several species of algae, which are attached either to other species of algae or to marsh grasses.

The animals of estuaries and related wetlands are tremendously important, not only as denizens of their environments but also for their role in marine communities and in human economics.[10] The best-known estuarine animals are detritus feeders such as oysters, clams, lobsters, and crabs. Several insect larvae, annelid worms, and mollusks enter the estuary from fresh water; most nearshore marine zooplankton can also be found partway into the estuary, along with several types of larger animals. Most important, perhaps, estuaries are the nursery grounds for vast numbers of marine animals ranging from shrimps and crabs to fishes. For example, more than 90% of the marine fish harvested by American fishermen comes from the continental shelf. Of these, about two thirds either spend their larval years in estuarine nurseries or must pass through estuaries between spawning and adulthood.

The estuarine ecosystem is complex and significant. It is also vulnerable, since estuaries have served as conduits for shipping and as sites for cities throughout human history. Estuaries are inhabited by animals that are adapted to a changeable environment, to be sure, but their strategic location has led to human alterations to a substantially greater degree than in almost any other ecosystem. Many people look upon estuaries as areas whose greatest value is to be filled and built upon, or to serve as dumping grounds for garbage, sewage, and industrial wastes. But their tremendous productivity can be made to serve as a food source for people —indeed, it is already a very important food source in the Far East—and almost all the major marine fisheries of the world are totally dependent on the estuaries for their continuance.

Oceans

Much of what has been said about lakes also applies to the oceans, but with certain differences. Seawater is saline, with an average salinity of about 35‰ distributed among the ions shown in Table 7.2. Unlike fresh water, sea-

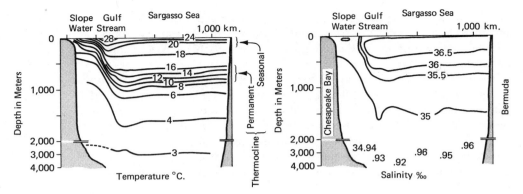

Figure 7.30
Distribution of temperature and salinity in a section of the ocean running between the island of Bermuda and Chesapeake Bay, crossing the Gulf Stream. Note that there is a considerable vertical exaggeration. [Reprinted, with permission, from G. L. Pickard, *Descriptive Physical Oceanography; An Introduction.* Copyright © 1963 by Pergamon Press Ltd., Oxford.]

water does not have a density maximum at 4°C; rather, it becomes continuously denser as it gets colder. As in lakes, the seas may be thermally stratified, with the thermocline determined mainly by the depth to which storms can mix the surface waters. However, the salinity of the sea is not uniform from top to bottom, and there may also be a pronounced salinity gradient (*halocline*) separating top and bottom waters. Figure 7.30 shows the vertical distribution of temperature and salinity in a representative area of the ocean. The great depth of the sea precludes any analog to the overturn experienced by lakes. Despite the lack of overturn, surface waters do mix with deep waters through the worldwide system of surface and bottom currents (Figure 3.17) and zones of upwelling (page 71).

Life Zones in the Ocean

The distribution of life zones in oceans is similar to that in lakes, except that the oceans are more complex and much deeper. A diagrammatic cross section of a portion of the ocean adjacent to a continent shows several different regions (Figure 7.31). The underwater extension of the continent is the *continental shelf*, which generally extends to a depth of roughly 125 to 200 m. The edge of the continental shelf may be close to the shore, or it may be several hundred kilometers offshore. From the edge of the continental shelf there is a more rapid descent, the *continental slope*, to the broad flat *abyssal*

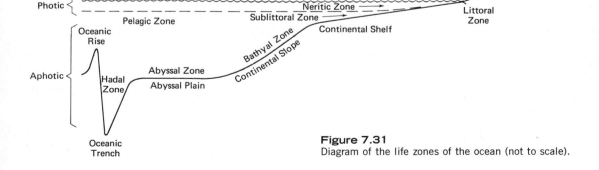

Figure 7.31
Diagram of the life zones of the ocean (not to scale).

plain that underlies most of the ocean at a depth of 4000 to 5000 m. From the abyssal plain rise great mountain ranges, the *midoceanic ridges*. Some of the isolated islands in the middle of the world's oceans (e.g., Hawaii, Iceland) are the tips of these ridges projecting above sea level. In addition, deep troughs drop down more than 11,000 m below sea level. Life exists from the surface of the sea to the bottom in virtually all areas, although its abundance is exceedingly variable.

Benthic life zones are defined in terms of the ocean's physical subdivisions (Figure 7.31). The *littoral*, or *intertidal* zone, is the zone between high tide and low tide levels. The *sublittoral* extends from the low tide mark to the edge of the continental shelf, with the *bathyal* zone comprising the continental slope. The *abyssal* zone includes the abyssal plains, and the *hadal* zone includes any life in the deep trenches below 5000 m. The open water from shore to the edge of the continental shelf is the *neritic* zone; the rest of the ocean is the *oceanic*, or *pelagic*, zone. Water above the compensation level is within the *photic* zone; the remainder of the sea is in the *aphotic* zone.

The open sea as a whole is not a productive place. The low level of productivity means that ocean waters are extraordinarily clear. The photic zone may range down to 150 m below the surface. In the clearest portions of the sea; it is about half that in more productive zones, and it is substantially less in nearshore waters containing sediment particles derived from land that have not yet settled out.

Marine Communities

Although life in the sea is not particularly abundant, the number of different types of organisms is exceedingly high. Every major group of algae and almost every major group of animals can be found somewhere in the oceans. The only striking omissions are the vascular plants and insects, which are abundant in estuaries but which have few marine representatives.

The most abundant pelagic phytoplankton are diatoms and dinoflagellates, which together produce most of the organic carbon in the sea (and most of the oxygen in the atmosphere), as well as other forms of golden-brown algae and flagellated green algae. Some seaweeds, such as the large brown alga *Sargassum*, may have a floating stage. Pelagic zooplankton include representatives of every major group and most minor groups, either as permanent members of the plankton community or as transients during their larval stages. Common among permanent planktonic forms are the complex and beautiful protozoa known as foraminifera and radiolaria, arrow worms (*Sagitta*), certain annelid worms, swimming snails, jellyfish, and, most abundant of all, the crustacea such as shrimp and copepods. Among the temporary zooplankton are the larvae or juveniles of the larger animals from all marine environments and even some freshwater environments.

The largest animals in the open ocean are the swimmers. Squid and nautili are the most prominent invertebrates, and marine vertebrates include bony fishes, sharks, sea turtles, and mammals such as seals and whales. Sea birds also feed on many of the same food types as these large carnivores. Air-breathing swimmers like the turtles and whales are found mainly in the photic zone, but fishes extend from the surface to the bottom. We are all well acquainted with the fishes that live near the surface; they include tuna, shark, sardine, mackerel, herring, bonito, and anchovy. Those of greater

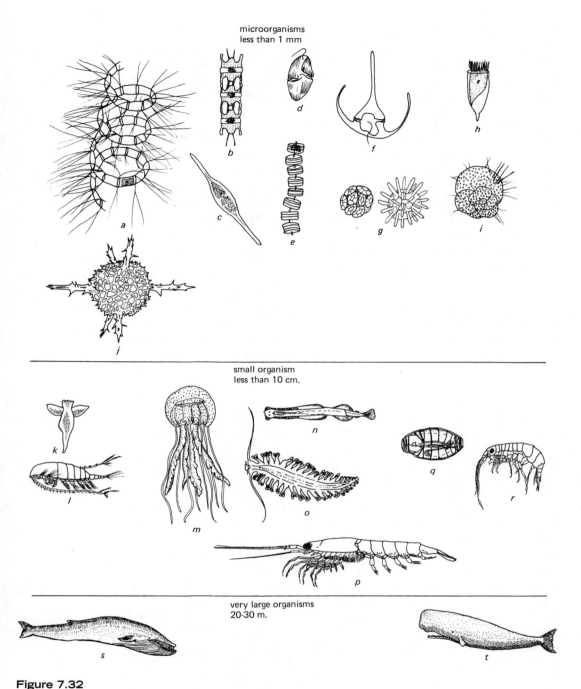

microorganisms
less than 1 mm

small organism
less than 10 cm.

very large organisms
20-30 m.

Figure 7.32
Organisms found in the pelagic zone of the ocean: (*a*) *Chaetoceros,* a colonial diatom; (*b*) *Biddulphia,* a colonial diatom; (*c*) *Nitzschia,* a diatom; (*d*) *Gymnodinium,* a dinoflagellate (one species of this genus is the cause of the highly toxic "red tide"); (*e*) *Thalassiosira,* a colonial diatom; (*f*) *Ceratium,* a dinoflagellate; (*g*) a pair of coccolitho-phorids, very small algae with calcareous plates; (*h*) *Favella,* a tintinnid, a protozoan with a chitinous cup surrounding the body, (*i*) *Globigerina,* a foraminifer, an amoeboid protozoon with a multichambered calcereous test; (*j*) *Protocystis,* a radiolarian, an amoeboid protozoon with a complex silica test; (*k*) *Clione,* a shell-less pteropod snail; (*l*) *Calanus,* a copepod crustacean; (*m*) *Pelagia,* a jellyfish medusa; (*o*) *Tomopteris,* an annelid worm; (*n*) *Saggita,* the arrow-worm; (*p*) *Euphausia,* an euphausiid crustacean, commonly known as "krill" and serving as the main food for whalebone whales; (*q*) *Cyclosalpa,* a tunicate; (*r*) *Apherusa,* an amphipod; (*s*) *Balaenoptera,* the blue whale, a whalebone whale; (*t*) *Physeter,* the sperm whale, a toothed whale.

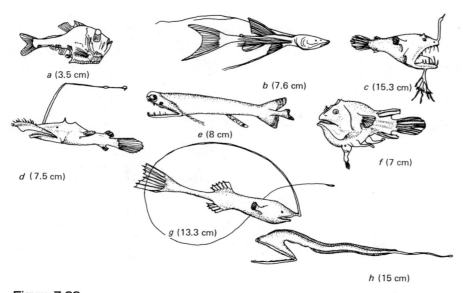

Figure 7.33
Fishes of the deep waters of the ocean: (a) *Argyropelecus*, 300 m. depth; (b) *Bathypter-ois*, 550 m. depth; (c) *Linophryne*, 1,500 m. depth; (d) *Lasiognathus*, 2,000 m. depth; (e) *Malacostus*, 100–2,500 m. depth; (f) *Edriolynchus*, 2,000 m. depth; (g) *Gigantactis*, 2,500 m. depth; (h) *Macropharynx*; 3,500 m. depth. [From H. U. Sverdrup, Martin W. Johnson, and Richard H. Fleming, *The Oceans: Their Physics, Chemistry, and General Biology*. Copyright © 1942, renewed 1970. Reprinted by permission of Prentice-Hall, Inc., Englewood Cliffs, New Jersey.]

depth are often grotesque and unlike any fish found at the surface. They tend to be small—15 to 20 cm. is large for the ocean deeps—and they are exceedingly dispersed. Many have luminescent appendages with which to lure prey, or mouths that look several sizes too large for the rest of the body. Food is not plentiful in the deep waters of the sea, so these fishes must go for long periods of time without food and then consume as much as they can when they have a chance.

The deep ocean benthos is diverse. Food is a little more plentiful on the bottom than it is in the deep waters of the ocean, since the bottom is the ultimate resting place for all detritus raining out of the upper layers. Sea cucumbers, brittle stars, crinoids (sea lilies), sea urchins, certain benthic fishes, and several types of crustaceans, as well as sea anemones, clams, and similar animals are all found on the bottom, not only on the abyssal plains but some even at the bottoms of the deepest ocean trenches. Many are detritus feeders, but many are carnivores. In fact, the biomass of carnivorous brittle stars is often higher than that of the detritus feeders that serve as their food source; however, their metabolic rate is much lower, so their assimilation is less. The diversity of animals on the ocean bottoms may be considerably higher than that of the overlying communities of the photic zone, even beneath relatively sterile tropical waters. It has been suggested that this diversity reflects the great constancy in the physical environment of the deep sea.

The communities of the continental shelf are both richer and more diverse than those of the open ocean. Diatoms and dinoflagellates are still the most productive phytoplankton, but green, brown, and red algae anchored to the bottom in the shallower regions may be significant. Some of the great kelps

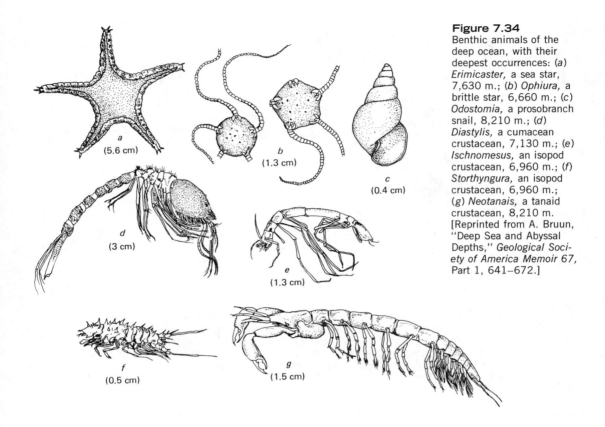

Figure 7.34
Benthic animals of the deep ocean, with their deepest occurrences: (*a*) *Erimicaster*, a sea star, 7,630 m.; (*b*) *Ophiura*, a brittle star, 6,660 m.; (*c*) *Odostomia*, a prosobranch snail, 8,210 m.; (*d*) *Diastylis*, a cumacean crustacean, 7,130 m.; (*e*) *Ischnomesus*, an isopod crustacean, 6,960 m.; (*f*) *Storthyngura*, an isopod crustacean, 6,960 m.; (*g*) *Neotanais*, a tanaid crustacean, 8,210 m. [Reprinted from A. Bruun, "Deep Sea and Abyssal Depths," *Geological Society of America Memoir 67*, Part 1, 641–672.]

of the northern rocky shores are even harvested commercially in the United States, and several types of seaweeds are harvested as human food in the Far East. The zooplankton are generally similar to those of the pelagic region, but some purely open-sea species have been replaced by nearshore species, and the overall diversity is higher. Temporary zooplankton are more abundant over the continental shelf than in the open ocean.

The swimmers in the oceans over the continental shelf are diverse, and many are well known, for they include almost all economically important fish species, as well as large squid, whales, seals, sea otters, and sea snakes. The most numerous are the fishes, which include many species of shark among their ranks, as well as herringlike species (e.g., menhaden, herring, sardine, and anchovy), cods and their relatives (e.g., haddock and pollack), salmon and sea trout, flatfish (e.g., flounder, sole, plaice, and halibut), and mackerels, including tuna and bonito.

The benthos of the continental shelf includes a wide variety of animals, among them snails, clams, shrimp, lobster, crabs, sea cucumbers, sea urchins, brittle stars, starfish, sponges, anemones, bryozoa, annelid worms, and foraminifera. These assemblages are more variable than those of deeper waters, since the physical factors affecting the shelf are themselves more variable. The bottom may be mud, sand, or rock, and there is a greater temperature differential in the sublittoral than in deeper ecosystems. The benthic community reflects these differences. Many organisms live in the bottom sediments, although the generic makeup of any particular community is controlled largely by the nature of the substrate. The assemblages of

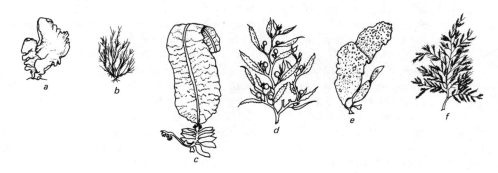

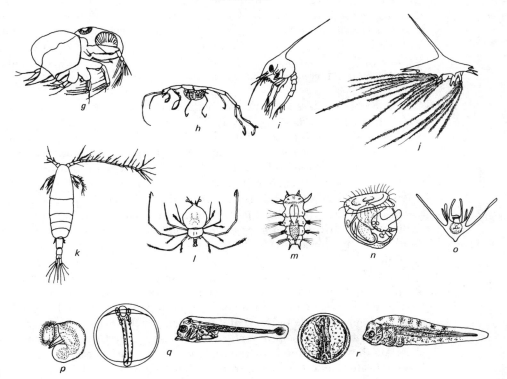

small animals
less than 1 cm.

Figure 7.35
Organisms of the neritic zone of the ocean: (a) *Ulva,* the sea lettuce, an attached green alga; (b) *Ectocarpus,* an attached brown alga; (c) *Alaria,* a broad-leafed attached brown alga; (d) *Sargassum,* a brown alga; (e) *Palmaria,* a broad-leafed red alga; (f) *Polysiphonia,* an attached red alga; (g) *Podon,* a carnivorous cladoceran; (h) *Phtisica,* an amphipod; (i) *Thia* larva, the planktonic stage of a crab; (j) Barnacle nauplius, an early pre-adult stage; (k) *Acartia,* a copepod; (l) Phyllosoma larva of a lobster; (m) *Platynereis* larva, a polychaete worm; (n) *Ostrea* larva, the common oyster; (p) Snail larva; (o) Brittle star larva; (q) *Gadus* (haddock) egg and larva; (r) *Solea* (sole) egg and larva.

animals living on the surface are even more variable, especially in the zone where light penetrates to the bottom. In these areas, temperature may be a very important factor. Benthic assemblages are best developed in tropical areas and are practically absent in the Arctic and Antarctic.

The intertidal zone is the most variable zone in the entire sea. It is completely covered at high tide and completely uncovered at low tide except for tide pools. Any organism that can survive in the intertidal must be either resistant to periodic desiccation or able to burrow to water level. Like the estuary, the intertidal is a zone of very high productivity within a simple community, many of whose members may be exceedingly abundant. For instance, in a sandy beach in California, clam diggers removed over 2000 tons of Pismo clams (*Tivela stultorum*) in two and one-half months. More than 100,000 individuals of certain annelid worms can be found per cubic meter.[11]

There is no typical littoral zone: a rocky intertidal is different from a sandy beach or a muddy intertidal. All have some things in common, however. The wave action is stronger here than anywhere else in the sea. The substrate erodes rapidly, and turbidity is high. There are few species of plants. Those that are present are firmly anchored to the substrate, and they may be very abundant. The animal community depends for its energy base on abundant detritus washed in by the waves. The most common animals are clams, snails, barnacles and other types of crustacea, annelid worms, sea anemones, and sea urchins. There is a conspicuous zonation with respect to the tide (Figure 7.36), in which animals that are more resistant to desiccation tend to occur at higher levels than those that are less resistant.

CORAL REEFS. There are other specialized ecosystems in the ocean. *Coral reefs* are among the most significant. These are among the most productive of all ecosystems anywhere, with a diversity equalled only by tropical rainforests. The reef environment is subject to the constant pounding of the waves, and parts of the reef are often above water at low tide. Wave erosion is so great that reefs are constantly being torn apart, especially in areas where the reef-building organisms have died. But this constant pounding is essential for the survival of the reef as an environment. Efforts to create artificial reefs in the laboratory ended in failure until recently, when the Smithsonian Institution in Washington built a "reef" around a wave-making machine.

The reef is a very rich environment for those species that can withstand the physical pounding of the water. Oxygen concentrations are very high. During the day, they may reach 250% of saturation, because of the production of oxygen by algae in the reef itself. Many of these algae live symbiotically within the tissues of the reef corals. Likewise, the tremendous amount of carbon dioxide produced by the reef metabolism can be carried off before it becomes too high. It is not surprising that reefs continue to grow in some places even as the waves wear them down in others.

Reefs can be found wherever a suitable substrate exists in the lighted waters of the tropics, away from areas of continental sediments that would tend to bury the reef in mud, or cold upwellings that would lower the temperature too much. *Barrier reefs* form adjacent to continents, *atolls* build up on top of submarine heights, and several intermediate types exist. In all modern reefs, the main reef structure is formed by corals and calcareous red

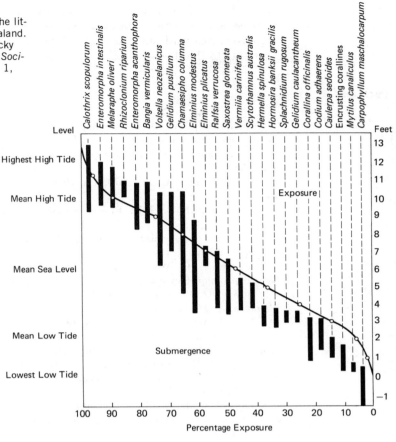

Figure 7.36
Vertical zonation of animals of the littoral zone at Auckland, New Zealand. [Redrawn from M. S. Doty, "Rocky Intertidal Surfaces," *Geological Society of America Memoir 67*, part 1, 535–586.]

and green algae, both of which precipitate calcium carbonate skeletons. Although the name "coral reef" implies that corals are the most abundant organism, algae of many sorts constitute over three times the biomass of the corals. In addition, the calcareous algae are not only producers; their skeletons cement the entire reef to make it more resistant to the pounding surf. Conditions vary greatly from one part of a reef to another: High in the fore-reef, where surf action is strongest and diurnal temperature variation is about 3°C, corals are most abundant; in the quieter area of the back-reefs, where oxygen concentration is slightly lower and diurnal variation is as much as 12.5°C, corals may be absent.

In addition to corals and calcareous algae, foraminifera and mollusks are important elements in the total reef structure. Sponges, worms, sea urchins, and fishes are also both abundant and diverse. The great productivity of reefs is obviously due to more than the abundance of light and oxygen. These are abundant in other areas of the sea, also, but the productivity in these other areas is limited by nutrient availability. Nutrients cannot rain out of the lighted zones of a reef and be lost, as they can in the pelagic zone. In addition, the incredible diversity of the reef community has evolved in such a way that nutrients are recycled efficiently within the reef, and virtually no organic matter is incorporated permanently into the reef structure. It is possible that some nutrients may be suboptimal for some reef organisms, but rapid nutrient cycling provides a mechanism to circumvent the usual influ-

AQUATIC
ECOSYSTEMS

209

ences of the ocean's natural paucity of nutrients. It may also allow a gradual increase in nutrient availability by incorporating the nutrients of the nonreef organisms that wash into the reef, while limiting the loss of the reef's own nutrients to the deeper seas.

Notes

[1] Tilman, D., 1977. Resource competition between planktonic algae: An experimental and theoretical approach. *Ecology*, **58**, 338–348; Tilman, D., 1980. Resources: a graphical-mechanistic approach to competition and predation. *Am. Nat.*, **116**, 362–393.

[2] Russell-Hunter, W. D., 1970. *Aquatic Productivity*. New York: Macmillan Publishing Co.

[3] Wielgolaski, F. E., 1975. Biological indicators on pollution. *Urban Ecology*, **1**, 63–79.

[4] 1 acre-foot is equivalent to a volume of water 1 acre in area by 1 foot deep. An acre is $43,560 \text{ ft}^2$, so that an acre-foot is $43,560 \text{ ft}^3$, or 1233.5 m^3.

[5] Leopold, L. B., Wolman, M. G., and Miller, J. P., 1964. *Fluvial Processes in Geomorphology*. San Francisco: W. H. Freeman and Company.

[6] McRoy, C. P., and Barsdate, R. J., 1970. Phosphate absorption in eelgrass. *Limnology and Oceanography*, **15**, 6–13; McRoy, C. P., Barsdate, R. J., and Nebert, M., 1972. Phosphorus cycling in an eelgrass (*Zostera marina*) ecosystem. *Limnol. and Oceanog.*, **17**, 58–67; Bristow, J. M., and Whitcombe, M., 1971. The role of roots in the nutrition of aquatic vascular plants. *Am. Jour. Botany*, **58**, 8–13.

[7] Chapin, J. D., and Uttormark, P. D., 1973. Atmospheric contribution of nitrogen and phosphorus. University of Wisconsin Water Research Center Technical Report, **WIS-WPC-73-2**.

[8] National Academy of Sciences, 1969. *Eutrophication: Causes, Consequences, Correctives*. Washington, D.C.: National Academy of Sciences, section on "geographic concepts."

[9] Hutchinson, G. E., 1969. Eutrophication, past and present. In National Academy of Sciences, 1969. *Eutrophication: Causes, Consequences, Correctives*. Washington, D.C.: National Academy of Sciences, 17–26.

[10] Gosselink, J., 1980. *Tidal Marshes: The Boundary Between Land and Ocean*. Washington: U.S. Fish and Wildlife Service.

[11] Hedgpeth, J. W., ed., 1957. *Treatise on Marine Ecology and Paleoecology*, vol. 1: Ecology. Geological Society of America Memoir 67, pt. 1.

Terrestrial Ecosystems

8

ALTHOUGH ONLY ABOUT a quarter of the earth's surface is dry land, the complexity and variety of terrestrial ecosystems are greater than those of aquatic ecosystems. Terrestrial ecosystems have no unifying theme analogous to the physical properties of water. Rather, the variety of climates, the diversity of soils, and the heterogeneity of landscape all conspire to give rise to a wide variety of themes that are woven together in nature to give an incalculable diversity of communities and ecosystems.

An aquatic ecosystem is essentially a single-phase system, where water sets the tone for the entire habitat. A terrestrial ecosystem, on the contrary, is a three-phase system, where the characteristics of the habitat are a function of the atmosphere and climate, the soil, and the biotic community itself. The atmosphere is the source of oxygen for animals and carbon dioxide for plants. Oxygen is sufficiently common and well mixed in the atmosphere that it does not limit growth, and carbon dioxide cycles efficiently enough that it seldom limits plants, despite its scarcity in the atmosphere. Air as a medium of support is much less buoyant than water; therefore an organism needs a rigid skeleton to live and move on land. Climate is much more variable than in aquatic ecosystems. Rainfall is never totally predictable, even in the most uniform of environments, and lack of water may be a source of extreme stress in arid regions or during periods of drought. Temperatures are much more variable on land than in water, since they are controlled by the absorption of heat by soil, rocks, and vegetation. A given quantity of heat energy absorbed by these materials can change their temperature two to five times as much as the same energy absorbed by water. Thus, heat gained and lost by terrestrial ecosystems results in a wide fluctuation of temperature, both diurnally and seasonally.

The soil serves two major functions: It provides support for living organisms, and it is the source of all essential nutrients except for carbon, oxygen, and hydrogen. The role of soil as a source for nutrients is unique to terrestrial environments. It is the site of the entire detritus food chain, and thus is central to the biogeochemical cycling of nutrient materials. Different types of soils have different properties that affect the availability of nutrients to plants, so that productivity of terrestrial ecosystems is closely tied to the chemistry of their soils.

Finally, there are very few aquatic ecosystems other than the coral reef in which living organisms leave a permanent mark on the entire ecosystem or are major factors in creating the habitat for other organisms. Both the diversity and the degree of influence of interspecific interactions are much more complex in terrestrial ecosystems than they are in aquatic ecosystems. For example, the habitat of a flower on a forest floor is as much a function of the forest trees as it is a function of the soils and climate, since the trees determine how much light reaches the flower. Plants have a role in breaking down and weathering rock and building soil that has few parallels in aquatic ecosystems. In addition, ecological succession in a terrestrial community is driven almost entirely by the organisms within the ecosystem, unlike typical aquatic successions (e.g., lake eutrophication), which depend largely on materials being washed in from outside.

Climate in Terrestrial Ecosystems

The climate of an ecosystem is its most prominent and significant physical environmental factor. More than any other, it determines the characteristics of the community, and it has a significant role in other key aspects of the abiotic environment as well. Climate is a complex phenomenon that cannot be described adequately by one or two factors alone. Its most important aspects are temperature and water relations. These include mean temperature and fluctuations around it, mean annual precipitation and its timing and predictability, and potential evapotranspiration.[1]

Temperature and precipitation are independent of one another. Mean annual temperature ranges from high to low, and temperature fluctuations range from extreme to minor. Precipitation in wet areas can be continuous or concentrated in monsoons, and rainfall in dry climates may be fairly evenly spread or concentrated into a single rainy season. If there is a rainy season, it may come during the winter or during the summer. Potential evapotranspiration depends on temperature, wind, and humidity.

Mean Temperature

Temperature influences all chemical reactions, both inorganic and biochemical. As a rule, an increase of 10°C results in a doubling to tripling of the reaction rate. The important effects in ecosystems include both physical processes such as rock weathering and the responses of living organisms. We can classify climates on the basis of their mean temperature, as shown in Table 8.1. These classes reflect the gross differences in heat absorption by different ecosystems, and they have proven both useful and meaningful.

The basis of climatic differences around the earth is the very uneven heating of the earth by the sun. Much more solar energy heats the equator than the poles, so that the gradient from tropical to very cold is primarily due to latitude. Mean annual temperature is not determined exclusively by the

Table 8.1 Classification of Climates by Mean Temperature

Based on the temperature portion of Holdridge's (1947) classification of climates.

Mean Annual Temperature (in °C.)	Climatic Class
1.5	Polar: Perpetual Ice and Snow
3	Frigid: Tundra
6	Cold
12	Cool Temperate
17	Warm Temperate
24	Subtropical: Winter Freezing Rare
	Tropical

latitude, however. Temperature patterns are altered by variations such as altitude and deflection of climatic zones by mountain ranges.

The mean temperature of an ecosystem provides only a first approximation of the biologically meaningful temperature of the environment. Most organisms become inactivated by cold about the freezing point of water, and they go into a heat coma about 40 to 50°C. But organisms living in ecosystems with variable temperatures have developed adaptations to withstand temperatures that would normally be beyond their tolerance. These include exceedingly resistant stages in their life cycles such as spores, seeds, or cysts. Some animals have resistance-providing behavioral adaptations such as hibernation. In such cases, the mean temperature may have little relation to the range of temperatures during which the community is active and well developed.

Temperature Fluctuation

Temperatures are never constant all year long. They vary both diurnally and seasonally. But temperature fluctuations can be minor, or they can be extremely large. For example, the West African city of Agades, Niger, and the Central American city of Cristobal, Panama, have essentially the same annual mean daily temperature. But the variation about the mean throughout the year is some 14°C in Agades and about 1°C in Cristobal. The differences in the two ecosystems are enormous: Agades is on the edge of the Sahara Desert, and Cristobal is surrounded by tropical rainforest. Likewise, the South Island of New Zealand has a very low temperature fluctuation. Tree ferns and parrots, usually confined to the tropics and subtropics, are found within 5 to 6 km of glaciers, and palms come within 100 km. Parrots also exist in Tierra del Fuego, South America, whose mean annual temperature is similar to that of Halifax, Nova Scotia. Uniformity of temperature and the infrequency of frost imply a lack of stress that allows the survival of species that cannot withstand cold stress.

Rainfall

Few things are more important to living things than water. With very few exceptions, it is introduced into terrestrial ecosystems as rain. Plants utilize great quantities in their growth: a crop plant in the temperate zone transpires 200 to 500 g of water for every gram of plant tissue produced; the figure may be almost twice this in an arid region. Animals drink a great deal of water directly.

The distribution of rainfall over the earth is remarkably uneven. It varies from essentially zero in some dry areas such as the Sahara to well over 5000 mm/yr. It is not particularly meaningful to subdivide climates simply on the basis of rainfall classes, however, because plants and animals do not utilize rain per se; they use the water found in the soil. This quantity is a function not only of rainfall, but of the tendency of water to be lost through evapotranspiration, runoff, and percolation into the groundwater system below the root zone.

Fluctuation of Rainfall

Like temperature, rainfall is often seasonal. Even in areas where temperatures throughout the year are essentially invariant, the rainfall may be strongly seasonal. There are other ecosystems in which the opposite holds

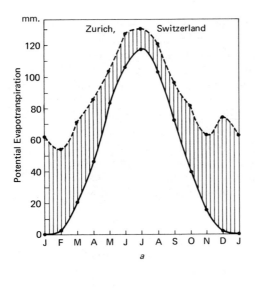

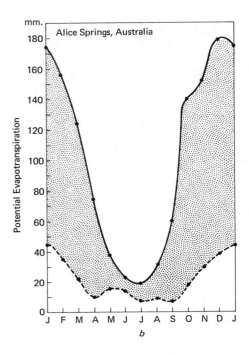

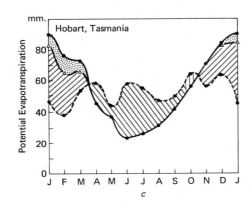

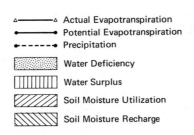

△———△ Actual Evapotranspiration
●———● Potential Evapotranspiration
●- - - -● Precipitation
Water Deficiency
Water Surplus
Soil Moisture Utilization
Soil Moisture Recharge

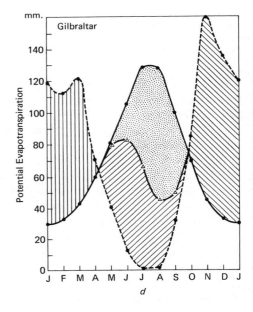

Figure 8.1
Water balance for selected stations. [Reprinted from D. B. Carter and J. R. Mather, "Climatic Classification for Environmental Biology," *C. W. Thornthwaite Associates, Publications in Climatology*, **19** (4), 305–395.]

true. In two different ecosystems with the same total annual rainfall and mean temperature, the communities may be quite different if the distribution of rainfall is different. For example, a monsoon area is a very different kind of place from a rainforest. Both may have the same total rainfall, but it is distributed uniformly in the rainforest, whereas it is strongly seasonal in the monsoon forest.

Potential Evapotranspiration

It was mentioned in Chapter 3 that about 70% of the water that falls as precipitation is lost through evapotranspiration. The amount of water involved makes evapotranspiration an important concept, but there is a big difference between *actual evapotranspiration*, the amount of water actually lost from the soil into the atmosphere, and *potential evapotranspiration*, the amount of water that would be lost under optimum soil moisture conditions. Potential evapotranspiration is in many ways the more useful concept, because it relates water loss to temperature and other relatively simple physical parameters such as wind and relative humidity. Actual evapotranspiration depends not only on these but also on the composition of the community, soil type, and the amount of water in the soil. It may be as high as the potential evapotranspiration, but it is often less.

The concept of potential evapotranspiration allows us to deal with water in ecosystems as it actually affects living organisms. If precipitation is greater than potential evapotranspiration, there is a net water surplus that can percolate through the soil into the groundwater system, be lost as surface runoff, or be stored for a time in the soil. If precipitation is so low that actual evapotranspiration is less than the potential evapotranspiration, there is a net water deficit, which may exert considerable stress on the community. Plants must draw on water stored in the soil, utilize adaptations that conserve water, or wilt.

The water balance of the soil may vary markedly throughout the year (Figure 8.1). The most extreme conditions shown in the figure are Zurich (Figure 8.1a), in which precipitation is always in excess of evapotranspiration, and at Alice Springs (Figure 8.1b), where there is always a water deficit. The pattern shown in most areas, however, is of alternating water surplus and deficit, as in Hobart (Figure 8.1c) and Gibraltar (Figure 8.1d). Potential evapotranspiration is highest in the summer, but precipitation may or may not be correlated with temperature. It is in Zurich; it is not in Gibraltar. There is no simple relationship between water surplus or deficit and either the temperature or the rainfall. For instance, Gibraltar has $1\frac{1}{4}$ times the rainfall of Hobart, yet its annual rainfall distribution is such that it also has a longer and more intense period of water deficit. Hobart and Zurich have very similar mean annual temperatures, yet the patterns of rainfall and temperature leave Hobart with a summer water deficit and Zurich with a continuous surplus.

Climate as a Biological Factor

How do you classify climate in a way that is meaningful to ecosystems, that includes enough detail to be useful, and that does not get bogged down in terminology or phenomena that are marginal to living things? Many ways have been proposed over the years, and the difficulty in accepting a classification is that the factors to which plants and animals actually respond—

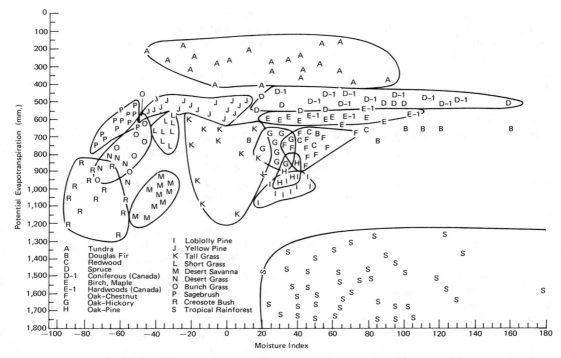

Figure 8.2
Relation between distribution of communities and climate, where climate is defined as potential evapotranspiration and a moisture index. The moisture index is calculated as follows:

$$\text{Moisture Index} = \frac{\text{Annual Water Surplus} - \text{Annual Water Deficit}}{\text{Annual Potential Evapotranspiration}} \times 100.$$

[Reprinted from D. B. Carter and J. R. Mather, "Climatic Classification for Environmental Biology," *C. W. Thornthwaite Associates, Publications in Climatology,* **19** (4), 305–395.]

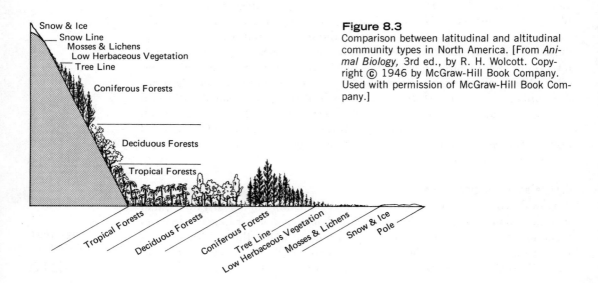

Figure 8.3
Comparison between latitudinal and altitudinal community types in North America. [From *Animal Biology,* 3rd ed., by R. H. Wolcott. Copyright © 1946 by McGraw-Hill Book Company. Used with permission of McGraw-Hill Book Company.]

habitat temperature, soil moisture, and available drinking water—are such a composite of the easily measurable parameters of temperature and rainfall. One cannot describe the relationship between communities and climate in terms of a climatic classification that includes all the climatic variables. No classification is entirely satisfactory.

One of the more useful is shown in Figure 8.2. It is based on potential evapotranspiration and a moisture index calculated from the magnitudes of the net annual water surplus or deficit. This approach combines mean temperature (as it relates to potential annual evapotranspiration with the annual rainfall) and the seasonal fluctuation of temperature and rainfall (as they are incorporated into the moisture index). However, aggregating climatic data over an entire year may conceal phenomena that influence the community to a considerable degree. Other methods of relating different aspects of climate to characteristics of terrestrial ecosystems are discussed by Carter and Mather.[2]

Nonclimatic Environmental Factors

The regional climate is the most obvious controlling factor for the structure of terrestrial ecosystems. But certain other factors are also significant, at least in some places. Of these nonclimatic factors, the most important deal with landforms, flooding, and fire.

Physical Landforms and Terrestrial Ecosystems

One of the first things we notice about any terrestrial ecosystem is the landform on which it is developed. Landforms influence not only the gross appearance of ecosystems, but also the existence of phenomena that mold microclimates and microhabitats. Different landforms are also characterized by different soils.

The characteristics of different landforms include *relief* (vertical distance from height of land to valley floor), amounts of *flatland* and *slopes*, and the relative prominence of erosion and deposition. Relief is analogous in many ways to position on the earth. Communities change with increasing altitude in much the same way they change going from the equator to the poles, since temperature drops as one climbs a mountain (Figure 8.3). Even in a tropical mountain area, the coldest climate at the top of the mountain may be similar to the arctic conditions of the far north.

Flatlands have a relief of less than about 150 m. They may be formed by erosion or deposition. They range from quite flat to gently rolling. Soils in flat uplands tend to be thin, but not as thin as on many slopes. Flat bottomlands tend to have thicker soils, and groundwater is plentiful. Percolation of water through the soil may be impeded in some places, and the soil may become locally waterlogged. The results of this will be examined in greater detail when we discuss the dynamics of soil formation.

Slopes are among the most significant and variable subunits of the terrestrial ecosystem. Erosion is common on slopes, but the rate is related to the steepness of the slope, its soil, and its vegetation. Steep slopes are unstable. Fewer plants can survive on steep slopes than in other environments, and those that can need extensive root systems to anchor themselves against landslides or eroding soil. Large animals of steep slopes are also uncommon;

in North America, only mountain sheep and mountain goats spring readily to mind. Low slopes, on the other hand, are among the most favorable environments for life. Although erosion removes the uppermost soil layers (at a rate averaging roughly 1 to 2 mm per decade, depending on the vegetation and the rainfall), weathering of the underlying rocks and accumulation of organic detritus can counteract the erosion and provide a relatively thick soil. In addition, a low slope tends to be well drained; excess water which cannot be held in the soil either percolates into deeper layers or runs off the surface. There are few pockets on a low slope in which water can collect to form a waterlogged soil.

The *aspect* of a slope, or the direction in which it dips, can have a profound effect on microenvironment. North-facing slopes in north temperate latitudes, for instance, never have the sun directly overhead, so radiation on them is oblique. If the slope is steep enough, it may never receive direct sunlight at all, and the only light is diffuse radiation. In arid and semiarid regions, the amount of water lost to evapotranspiration from a north-facing slope may be significantly less than that from a south-facing slope; for this reason, rolling hills in semiarid grasslands may be characterized by shrubs growing on all the north-facing slopes. Because they receive less solar energy, north-facing slopes are generally cooler than other slopes, and they may be covered by a cooler-climate community than that covering a south-facing slope. In very cold areas, north-facing slopes may be permanently frozen and barren, while south-facing slopes support some vegetation.

The dynamic interaction between the biota and the abiotic environment is shown in many ways in terrestrial ecosystems. We have touched on this subject before in conjunction with succession (page 146). It also appears with landform development. Water and wind moving over rocks and soil are a relentless force removing materials and shaping new landscapes. The most significant factor determining the rate of erosion is the health of the vegetation. Plants hold the soil and stabilize even a steep slope. A dynamic balance exists between the rate of erosion and the rate of soil formation on a healthy vegetated slope, and a soil in good shape promotes healthy plant growth. On a poorly vegetated slope, the rate of erosion tends to be much higher, and most or all of the soil can be lost.

There are natural mechanisms for destroying vegetation, including landslides and lightning-set fire. But people have been much more zealous than nature in removing vegetation through greatly increased burning, clearcutting of trees for timber, mining, and road building, and plowing of slopes for agriculture. These activities have resulted not only in soil destruction but also in extensive siltation into the waterways that receive the eroded soil. Furthermore, it is often difficult to revegetate a slope whose soil has been lost through erosion, and the careless destruction of slope vegetation may destroy the slope for productive purposes for a long period of time.

Flooding and Sediment Deposition

Many bottomlands are areas of sediment deposition. The most prominent are flood plains of rivers, which are built up from sediment that settle out during successive floods. This is the origin of most interior rolling plains such as the Great Plains of western North America, as well as the coastal plains adjacent to oceans. The communities that develop on flood plains are made up of species that can withstand periodic inundation of their root systems and produce root systems that effectively buttress the plant against falling over in the waterlogged soil during periods of flood.

If periodic flooding is halted for some reason, such as by construction of a flood-control dam, the balance of signals to which the community responds would be totally changed, and there would be major changes in community structure. A case in point is the alluvial flats redwood groves of California. Redwoods alternate with Douglas fir and oak hardwoods in the uplands, and they do not form pure stands as they do in the alluvial flats. Redwoods have thick, fire-resistant bark and the ability to regenerate their crowns quickly after a fire. In addition, they are well adapted to the unstable soils of the flats. They can form huge root buttresses to counteract the effects of uneven sinking during flooding, tolerate substantial burial by water and sediment, and send out new roots if the soil level changes through flooding.

None of the trees competing with redwood in the uplands is as resistant to the effects of both fire and flood. Thus, redwoods can form pure stands in the alluvial flats, where these are normal occurrences, and any competing species that might gain a foothold are removed by the normal flood cycle. Flood control dams have been suggested for the redwood region. Their construction would mean the removal of floods as a successional force on the alluvial flats, and trees like Douglas fir and oaks would be able to compete much more effectively with the redwoods in what is now their purest stands.[3]

Fire

The role of fire in natural ecosystems is often underestimated. Many fires are caused by lightning, and although people cause the great majority of modern fires, there is no question that fires have existed as long as there has been something to burn and lightning to start them. Most forests, at least in the temperate zones, have developed in concert with natural fires, and fire has a specific place in the forest's growth cycle. Indeed, many species depend on forest fires for their existence or propagation, and fire has had a key role in molding ecosystem structure since time immemorial.[4] Some plant species are more resistant to fire than others. They may have very thick bark, like the redwoods; they may have their actively growing portions protected either by an extensive tuft of leaves, as in the longleaf pine of the southeastern United States, or by burial in the ground; they may have reproductive organs that do not release their seeds until burned, as, for example, the cones of the jack pine.

Forest fire suppression was once the watchword of forest management. This affects forest ecosystems in several different ways. Certain branches of any tree are more shaded than others. They die, as do entire trees that are too shaded, diseased, defoliated, or weak. Dead wood accumulates on the forest floor. In a tropical rain forest dead wood can be broken down relatively quickly by the detritus food chain. In virtually no other ecosystem is this possible; the lignin that constitutes most of the tree's biomass is simply too tough and too difficult to break down during the season available to detritus organisms. Under normal circumstances periodic fires remove the dead wood but do not kill the healthy trees (Figure 8.4). Forest fire suppression does not diminish the rate at which fuel builds up. Under an extreme regimen of forest fire suppression, so much fuel accumulates that control is impossible if a fire ever does get started, and trees that would not normally be damaged by the fire are killed by it (Figure 8.5).[5]

a

b

Figure 8.4
Photographs *(a)* before and *(b)* after prescribed burning at Redwood Mountain Grove, King's Canyon National Park, California. Initially the forest had a great deal of combustible material and white fir saplings. The trash was consumed in the burn and the white fir saplings were killed, restoring the forest normal for the region. [Photographs courtesy Bruce M. Kilgore, U.S. National Park Service.]

Figure 8.5
Photograph of a 65-year Douglas fir
stand in the Gifford Pinchot National
Forest, Washington. Limbs are dead to
a height of 25 m., and self-pruning
has led to the accumulation of a great
deal of fuel on the forest floor. Should
a fire become started, it would be very
destructive and difficult to control.
(Compare with Figure 8.4a.) [Photo-
graph courtesy U.S. Forest Service.]

Of course, fire does a great deal more than consume fuel. It changes the
physical soil in several ways. For one thing, woody debris on the forest floor
contains considerable amounts of nutrient minerals that would be released
for recycling only very slowly through decomposition without fire. Fire releases them to the environment virtually immediately. Nitrogenous materials may be released either into the soil or into the atmosphere, depending on
the intensity of the fire. Other nutrients are concentrated in the ashes on the
forest floor.[6]

The rich ashy residue is an excellent seed bed for many species of trees.
Some species depend on it and their reproduction is vastly decreased without it. One example is the "big tree" (*Sequoiadendron giganteum*) of the
Sierra Nevada. Table 8.2 shows the number of seedling sequoias in three
different burned areas of the Sierra Nevada for the three years following a
prescribed burning on Redwood Mountain in Kings Canyon National Park.
The first year after the burning, the average seedling sequoia population was
over 54,000 per hectare. Not a single seedling sequoia could be found in an
unburned control plot in the same general area. Sequoia as a population
depends on fire to release nutrients and to prepare a seed bed for the population.

Few natural forest fires damage soils excessively. Some humus may be
burned, but the humus layer is not terribly flammable and it is augmented
by the ash from the burned debris. A very hot fire, on the other hand,
destroys the humus layer as well as the dead wood. This often brings a

Table 8.2
Sequoia Seedling Response to Prescribed Burning in 1969 on Redwood Mountain, King's Canyon National Park, California.

	Size (ha)	Mature Trees (No. Sequoia/ha)	Seedling Sequoia (No. per hectare)		
			1970	1971	1972
Burn 1	1.52	7.2	32,560	5,382	1,211
Burn 2	2.47	11.3	31,350	3,230	941
Burn 3	2.53	22.9	99,161	10,764	1,077
Total	6.52				
Mean		14.9	54,357	6,459	1,177
Control	2.14	14.5	0	0	0

After Kilgore, B. M. 1973. The ecological role of fire in Sequoia conifer forests: its application to national park management. *Jour. Quart. Research,* **3,** 496–513.

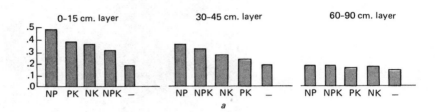

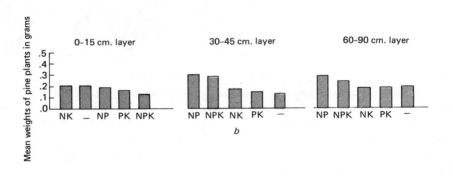

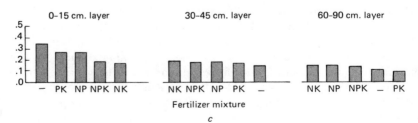

Figure 8.6
Bioassay showing loss of fertility from wildfire in ponderosa pine forests. Samples were taken from 3 levels of soil (0–15 cm., 30–45 cm., 60–90 cm.) and used as growth medium for ponderosa pine seedlings. Three different areas were used as the soil source. One *(a)* had no history of wildfire; the second *(b)* had been burned by wildfire 14 years before the study; the third *(c)* had been badly burned 3 years before the study. Mixtures of nitrogen (N), phosphorous (P), and potassium (K) were added to some of the soils; others were not fortified (—). [After Wagle and Kitchen, "Influence of Fire on Soil Nutrients in a Ponderosa Pine Type." *Ecology* 53: 118–125. © 1972, Ecological Society of America.]

pronounced drop in the soil's fertility and it may be very difficult for a forest to reestablish itself (Figure 8.6).[7]

Many natural communities are maintained by fire. Natural fires set by natural lightning have characterized the redwood forests of the Sierra Nevada from time immemorial, with the average period between fires on the order of about 9 years.[8] Trees such as the white fir would normally be killed while very young, since they are not nearly as fire resistant as the giant sequoia and sugar pine that make up the mixed sequoia-pine forests of the high Sierra. Young fir trees do not compete seriously with the sequoia or the sugar pine, but adult trees do if they become established. Fire suppression in the mountain areas inhabited by the giant sequoia have caused a marked proliferation of white fir in these forests, to the detriment of the natural environment of the sequoia forest and also to the sequoia itself (Figure 8.7).[9] Similar phenomena can be seen in other forests throughout the world.

Widespread communities such as the tropical grassland savannas of Africa owe their existence to fire, and they would be very different from what they now are if fire were not a normal part of the ecosystem. When fire is carefully excluded from the savanna, trees invade the area and displace grasses as the climax vegetation. The mechanism by which this occurs is shown in Figure 8.8. In the absence of fire, the vigorously growing woody plants proliferate and eventually become dominant in the community. Periodic fire burns both grass and woody species back to the ground, but the most actively growing part of the grass is at or just beneath the surface, where it is seldom harmed by fire, whereas the most actively growing portions of the woody species are destroyed by fire. In addition, the germination of many species of grass is stimulated either by the heat of the fire or by changes in the environment that result from it. Thus grasses can regenerate much faster after a fire than can woody plants.

There are many other examples. The Douglas fir forests of the northern Rocky Mountains will pass from Douglas fir to cedar-hemlock if fire is excluded, because Douglas fir seedlings cannot grow in the shade of the adults and those of cedar and hemlock can. However, if a fire sweeps through one of these forests, the cedar and hemlock will be burned out, and Douglas fir seedlings will spring up rapidly on the ash beds. An excellent review of the role of fire in terrestrial ecosystems is given by Ahlgren and Ahlgren.[10]

Interactions Between Plants, Soil, and Water

The most important interface in many ways between the living and nonliving parts of a terrestrial ecosystem is the root system of plants in the soil. All nutrient materials pass from the abiotic environment into the food chains at this point, as does the water that contributes to the carbohydrates fixed through photosynthesis. The relations between plants, soils, and water are very complex, but they are so fundamental to the functioning of terrestrial ecosystems that we must at least touch on them at this point.

Water, Soils, and Plants

Water has several different forms in soils. It may be chemically combined or uncombined. The most important types of chemically bound water are the water of crystallization of mineral grains and the water of hydration of clay

a

b

Figure 8.7
Two photographs of the Confederate Grove in Mariposa Grove, Yosemite National Park, California, showing changes in forest composition following fire suppression in the forest: *(a)* 1890, before any fire suppression policy; *(b)* 1970, after almost 50 years of conscious fire suppression. [The 1890 photo by J. J. Reichel, courtesy of Mrs. Dorothy Whitener; historical documentation by Bill and Mary Hood; 1970 photo by Dan Taylor. Both photos courtesy of U.S. National Park Service.]

Figure 8.8
Mechanism of fire control of a grassland climax: *(a)* mesquite is normally part of a grassland community; *(b)* in the lack of fire, it grows aggressively and may become dominant over grass; *(c)* fire reduces grass and mesquite alike, but the growing points of grass lie near or beneath ground, protected, while those of mesquite are exposed; *(d)* grass recovers quickly, while mesquite loses several years' growth. [From "The Ecology of Fire," by C. F. Cooper. Copyright © 1961 by Scientific American, Inc. All rights reserved.]

mineral particles. These affect certain properties of soils, but they are not at all available to living organisms. Liquid water may be tightly or loosely held in the soil by *adhesion* (attraction of water molecules to solid particles) and *cohesion* (attraction of water molecules to each other). The tension by which water is held in the soil depends on the amount of water present.

It makes sense to differentiate three types of soil water on the basis of the tension binding them to the soil: *hygroscopic, capillary,* and *gravitational.* Hygroscopic water is held so tightly that it shows few properties of liquid water and does not move except in the vapor phase. Capillary water is held in place in the small pores of the soil by moderate surface tension. Capillary water can move to a limited extent, and it is the major source of water for vascular plants. But only some capillary water is so loosely bound that it can be extracted by plants. Gravitational water moves freely through the soil under the influence of gravity. It fills the large pores during periods of precipitation, but it commonly drains soon thereafter. As a result, gravitational water takes up space that is usually filled with air. Most plants require an aerated soil for proper root metabolism, so that drainage of excess gravitational water benefits plant growth. In areas of high rainfall, soluble materials dissolve in the excess gravitational water as it percolates into lower levels. The removal of mineral nutrients by percolating gravitational water is known as *leaching.* It is a basic formative agent of many soils.

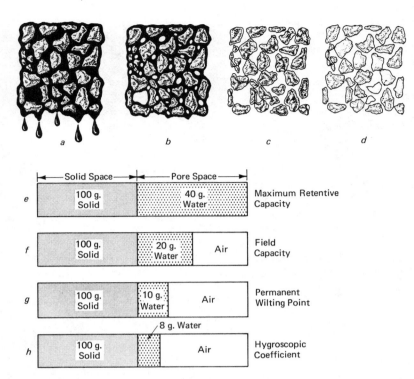

Figure 8.9
Illustrations of the water content of a soil show it at saturation (a,e), at field capacity (b,f), at its permanent wilting point (c,g), and in a state of extreme dessication, the hygroscopic coefficient (d,h). [Modified, with permission, from H. O. Buckman and N. C. Brady, *The Nature and Properties of Soils,* 7th ed. Copyright © 1969 by Macmillan Publishing Co., Inc., New York.]

Soils can hold only so much water. They reach their maximum retentive capacity when water fills all the pores between soil particles and there is no air space; this is *saturation* (Figure 8.9a). Any more water added to the soil runs off on the surface. When water is allowed to drain from a saturated soil, the gravitational water is lost and the larger pores refill with air, leaving only the small pores filled with capillary water. The amount of water retained in a soil by capillary attraction when drainage has reached equilibrium with retention is termed the *field capacity* (Figure 8.9b). Plants have limited abilities to extract water from soils. When they are growing in soils that have dried out so that water is held at a greater tension than their ability to extract it, they are no longer able to take up water, and they wilt. The tension equal to plants' abilities to take up water is referred to as the *permanent wilting point* (Figure 8.9c). If the soil dries out still further, so that all capillary water is lost from even the smallest pores leaving only hygroscopic water, the amount of water remaining in the soil is referred to as the *hygroscopic coefficient* (Figure 8.9d).

Water surplus and water deficit are functionally related to the interaction between plants, soils, and water. A water surplus is soil water in excess of field capacity. Because the pore space in most soils does not depend on water content, a water surplus means that pores normally filled with air are filled with water and drainage of gravitational water through the soil will result in at least some leaching. Impeding drainage for some reason (such as by a claypan layer close to the surface) means that pores cannot fill with air, and the oxygen content is much less than that of a soil at field capacity. Extensive leaching can remove the greatest part of the nutrient base from the upper soil horizons containing most plant roots.

Soil-water utilization is the removal of easily obtainable capillary water

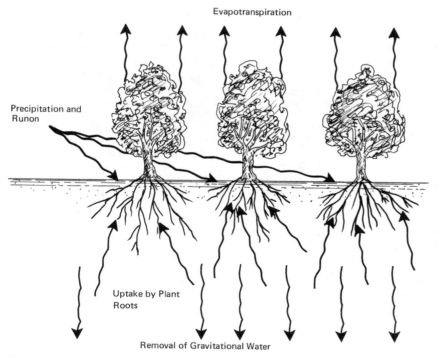

Evapotranspiration

Precipitation and Runon

Uptake by Plant Roots

Removal of Gravitational Water

Figure 8.10
Balance of water in the soil depends on water introduced by precipitation and runon and removed by gravity and evapotranspiration.

from soils by evapotranspiration in plants; soil-water recharge is its replacement back to field capacity. A water deficit is reached when the soil water content gets to the vicinity of the permanent wilting point. The balance of water in the soil depends on the interplay of the forces of drainage by gravity, evapotranspiration by plants, and recharge from precipitation (Figure 8.10). The optimal conditions for growth of most plants are realized when this balance holds soil water near (but not quite at) the field capacity. Since levels of soil water always oscillate in all soils, this minimizes both leaching and water deficits. When water supplies are greater than this, leaching is greater than optimal, and aeration is less; when water supplies are much less than this, water stress can result.

Soils

Soils are the most characteristic features of terrestrial environments. They are complex mixtures of weathered rock materials and organic detritus which are formed through the interacting physical, chemical, and biological processes occurring in the ecosystem. A soil is far more than "just dirt." It is the key to nutrient cycling in terrestrial ecosystems and the reservoir for nutrients and water, the medium for the detritus food chain, and the physical foundation for animals and plants. Its roles vis-a-vis nutrients and detritivores stem from its unique physical and chemical properties. Its physical role is due to the fact that plants are anchored to the soil by their root systems, and animals walk upon it and are supported by it.

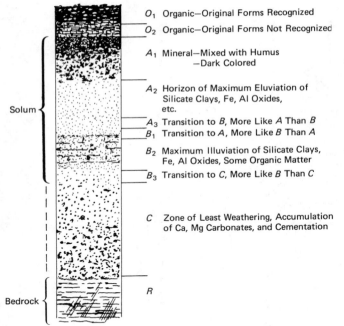

O₁ Organic—Original Forms Recognized

O₂ Organic—Original Forms Not Recognized

A₁ Mineral—Mixed with Humus
—Dark Colored

A₂ Horizon of Maximum Eluviation of
Silicate Clays, Fe, Al Oxides,
etc.

A₃ Transition to B, More Like A Than B

B₁ Transition to A, More Like B Than A

B₂ Maximum Illuviation of Silicate Clays,
Fe, Al Oxides, Some Organic Matter

B₃ Transition to C, More Like B Than C

Solum

C   Zone of Least Weathering, Accumulation
of Ca, Mg Carbonates, and Cementation

Bedrock

R

Figure 8.11
Generalized soil profile showing major horizons which may be present. Not all horizons are present in all soils, and some soils may be somewhat more complex. [Redrawn, with permission, from H. O. Buckman and N. C. Brady, *The Nature and Properties of Soils*, 7th ed. Copyright © 1969 by Macmillan Publishing Co., Inc., New York.]

Soil Profile

The way to begin to look at soils is in terms of the *soil profile*. Soils are typically composed of several layers lying parallel to the surface, which form a recognizable pattern when a top-to-bottom section is made. The soil profile is this cross-section. It reflects the climatic regimen under which it was formed, the problems in growing crops on it, and the deficiencies for which organisms in the associated ecosystems must be adapted. A typical soil profile has 5 main horizons and several transition levels. These horizons are commonly identified by a series of letters with subscripts (Figure 8.11).

The litter zone, or *O* horizon, consists of partially decomposed leaf litter. It can be differentiated into the O_1 horizon, which consists of fresh litter in which the original forms of plant and animal residues can be recognized with the naked eye, and the underlying O_2 horizon, which contains blackened and unrecognizably decomposed organic material. The O_1 level is thickest in a deciduous forest immediately after leaf fall, when the forest floor is covered with fresh leaves, and it is virtually gone by the end of the following summer, when the leaves have largely decomposed.

Underlying the litter zone is the *A* horizon, or *topsoil*. It is the site of intense biological and strictly physical activity. Many of the decomposers are located in the topsoil, and this is the horizon in which materials are brought into aqueous suspension or solution and are leached downward through the soil. There is a net movement of material out of the topsoil, but precisely what kinds of nutrients or other materials are affected and how much is leached depends on the chemistry of the soil and the amount of percolating gravitational water. The topsoil has three subzones. The zone of humus incorporation (A_1) is almost always dark colored and relatively rich in organic materials, thoroughly mixed with the mineral soil. The zone of maximum leaching (A_2) is typically a light-colored horizon from which materials are being removed most quickly. The A_3 horizon is transitional to the subjacent B horizon.

The B horizon, or *subsoil*, is the horizon in which much of the material leached out of the topsoil is precipitated and enriched. This too can be divided into three zones, of which the B_1 and B_3 are transitional to the A and the C horizons, respectively, and the $B2$ is the zone of maximum precipitation of transported material.

Underlying the B horizon is the weathered rock or sediment that serves as the parent material for the mineral fraction of the soil. This C horizon is virtually lacking in organic materials. It may be underlain by unweathered bedrock, sometimes called the R horizon.

The relative thickness and significance of the major horizons is variable. Depending on the soil in question, a zone may be thick, thin, or even absent, and the characteristics of each zone vary widely from soil to soil. But the basic significance of the horizons is the same for all soils: the $A2$ horizon is the zone of maximum leaching, the C zone is weathered parent material, and so forth. the concept of the soil profile is thus very useful because it is a descriptive tool that can be related easily to the dynamic factors shaping soils, and by which different soils can be compared.

Soil Formation

Soil represents the dynamic interaction of living and nonliving components of the ecosystem to a degree found in few other parts of the ecosystem. This can be visualized by thinking of soils as mixtures of materials derived from two separate sources, each of which is altered and mixed by natural mechanisms. The sources are organic detritus and the parent geological substratum. Alterations in these are influenced both by inorganic weathering and by biological activity. In each case, the properties of the end products of the alteration are very different from their initial properties, and it is these end products that give soils their unique characteristics. The soil-forming environment is subject to many different kinds of influences. As ecosystems develop, the soil may change noticeably as part of the succession process. Stable ecosystems show a dynamic equilibrium between the factors forming the soil that allows it to remain more or less constant through time.

WEATHERING OF ROCKS. A "typical rock" contains several different kinds of minerals. These include quartz, the inert SiO_2 which is best known as beach sand, the silicate minerals that form naturally at the high crystallization temperatures of igneous and metamorphic rock, and clay minerals, which are tiny complex silicate and oxide minerals that form at the temperatures found at the earth's surface. High-temperature minerals are unstable at the temperatures of the biosphere, and they break down readily to form clays. The general reaction type might be summarized

$$\text{High-temperature mineral} + H_2O \longrightarrow \text{Clay mineral}$$

The precise reactions depend on factors such as temperature and moisture, the nature of exudates from plants that can react chemically with the minerals, and the chemistry of the parent material. Different species of clay mineral may be formed under different conditions of environmental weathering.

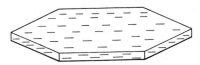

Figure 8.12
Diagram of a clay crystal. In the soil, clay surfaces are covered with a negative charge, and positively charged ions are adsorbed all over the surface.

Clay minerals are remarkable substances. They are tiny (0.01 to 5.0 μ)[11] broad flat crystals whose individual layers are loosely held, as in mica (which is related to clays), and may separate easily. Their surfaces carry an electric charge (Figure 8.12). This very small size, coupled with the flat shape of the crystals, gives them an extraordinarily large surface area; it has been estimated that the exposed surface area of clay particles in the top few inches of soil in a 2 hectare midwestern cornfield would be equivalent to the entire land surface area of the North American continent.

Clay minerals have certain properties that might not be anticipated *a priori:* They are highly reactive, and they are colloids. A *colloid* is a finely divided material that is dispersed through another medium. It shares many properties with a solution. For example, colloidal particles introduced into water tend not to settle out, and they can pass through very fine filters. In a soil, the colloidal clay particles can be carried by water percolating through the soil. Much of the leaching in topsoil involves the movement of clay colloids.

The negative charges on the surfaces of clay particles enables positively charged ions (*cations*) to bind to the surface of the clay with a weak chemical bond. This surface bonding is known as *adsorption*. Because the bonding is somewhat loose, cations adsorbed onto clay molecules can exchange places with other ions dissolved in the interstitial water (Figure 8.13). This process is known as *ion exchange*.

Because water moves so readily through soils, nutrients cannot persist in solution as in aquatic ecosystems. If there were no mechanism for nutrient storage in the solid fraction of the soil, the nutrients in the soil would quickly be leached out. Adsorption and ion exchange (Figure 8.14) provide a mechanism whereby they can be stored in a form that is available to living organisms. The acid exudates from plant roots release hydrogen ions that can replace nutrient cations on the clay particles. These can then be absorbed by the plant root through the soil water. If percolating gravitational water is slightly acidic (as, for example, in areas subject to acid rain), it can also supply hydrogen ions which replace other cations adsorbed onto the clay particles. The most important ion-exchange reactions can be summarized by Figure 8.15. The net results of the adsorption and exchange of nutrient ions on clay particles in soil are similar to those from the solution of nutrient ions in aquatic ecosystems. But as is typical of terrestrial systems, the precise mechanism is infinitely more complex.

```
Ca  H  H K  Mg  Ca              Ca  K H K  Mg  K K
 + +  +  + + +  +++               ++ + +  + ++  + +
 _ _  _  _ _ _  _ _ _             _ _ _ _  _ __  _ _
[      Clay Particle      ] + 5 KCl ⇌ [      Clay Particle      ] + 2 CaCl₂ + HCl
 _ _  _  _ _  _ _ _ _              _ _ _  _ _ __  _ _
 +  ++  ++ + +++                   +  ++ + + + + ++
 H  Mg  Ca H K  Ca                H  Mg  K K H K  Ca
```

Figure 8.13
Diagrammatic illustration of a portion of a clay particle showing exchange of cations adsorbed onto the surface. Addition of potassium ions into the soil solution causes exchange of potassium for other ions such as calcium and hydrogen, which are adsorbed onto the clay.

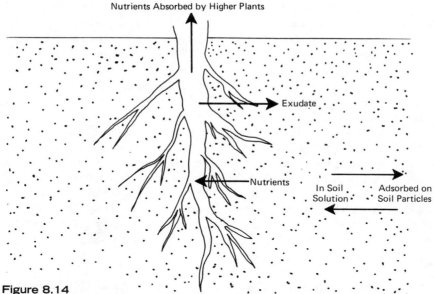

Nutrients Absorbed by Higher Plants

Exudate

Nutrients

In Soil Solution — Adsorbed on Soil Particles

Figure 8.14
Relationship between nutrients dissolved in soil water, adsorbed onto soil particles, and taken into the roots of higher plants.

Figure 8.15
Relationship between movement of nutrient cations (Nu+) in soils, cation exchange, and nutrient absorption by plants or leaching from soil.

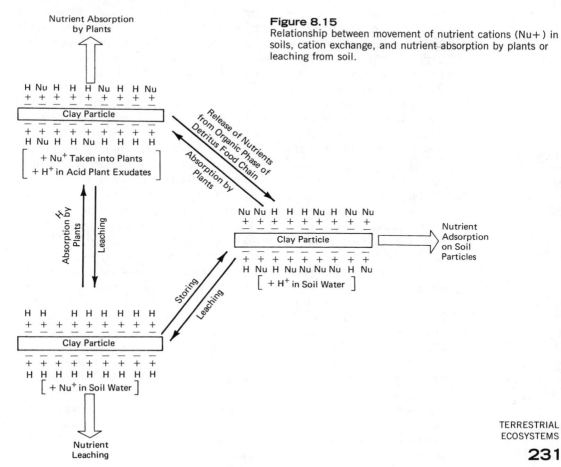

Nutrient Absorption by Plants

H Nu H H H Nu H H Nu
+ + + + + + + + +
— — — — — — — — —
Clay Particle
— — — — — — — — —
+ + + + + + + + +
H Nu H H Nu H H H H

[+ Nu$^+$ Taken into Plants]
[+ H$^+$ in Acid Plant Exudates]

Release of Nutrients from Organic Phase of Detritus Food Chain

Absorption by Plants

Absorption by Plants

Leaching

Nu Nu H H H Nu H Nu Nu
+ + + + + + + + +
— — — — — — — — —
Clay Particle
— — — — — — — — —
+ + + + + + + + +
H Nu H Nu Nu Nu Nu H Nu

[+ H$^+$ in Soil Water]

Nutrient Adsorption on Soil Particles

Storing

Leaching

H H   H H H H H H
+ + + + + + + + +
— — — — — — — — —
Clay Particle
— — — — — — — — —
+ + + + + + + + +
H H H H H H H H H

[+ Nu$^+$ in Soil Water]

Nutrient Leaching

FORMATION OF HUMUS. Organic material enters the soil as leaf litter, dead grasses and roots, animal wastes, and so on. It is broken down into smaller particles by the fungi, bacteria, and animals of the detritus food chain and mixed and aerated by the movement of the larger soil animals such as earthworms and soil-dwelling mammals. If conditions for decomposition were perfect, and if detritivore activity lasted long enough, all organic material would be completely broken down to carbon dioxide and water. This does happen in some cases, most notably in the wet tropics, but it is not common.

The end product of organic breakdown is typically a series of organic molecules collectively termed humic acids, or simply *humus*. The molecules are tiny, irregular in shape, and characterized by large surface area. They also bear a negative charge. Like the clays, they are colloids that have a very high ion exchange capacity. The ability of a typical humus particle to adsorb and exchange ions is 2 to 50 times higher than that of a typical clay. However, the clay content of most soils is so much higher than the organic content that the two colloidal fractions contribute roughly equally to the soil's ion exchange capacity.

MIXING OF THE ORGANIC AND MINERAL COMPONENTS. The distinctive layering of the soil profile results from the way in which the two major components are mixed. Mixing can be inorganic or organic. The most important inorganic agent is water. Percolating gravitational water carries along anything that has become dissolved in it, as well as colloidal clay and humus particles. Most of the material leached from the A horizon ends up in the B horizon, although some may be removed from the soil profile altogether. Different materials are precipitated at different distances from the surface in different soils. This accentuates the layering of the B horizon.

At the same time, the movement of the animals living in the soil (and there are vast numbers of them) stirs the soil mechanically. The most dramatic are doubtless the large burrowing vertebrates such as badgers and moles, but the activities of the smaller invertebrates account for greater soil mixing, especially in the A horizon. Much of the mixing is strictly mechanical, but there is also biochemical mixing, in which soil particles are ingested by the organisms. The inorganic particles are ground against one another in the animal's gut and mechanically abraded; they may even be chemically altered under the influence of digestive juices, bacteria, and fungi. The organic materials are broken down, yielding energy for the animal, and leaving castings that are rich in humic colloids and mineral nutrients.

The effect of different animals on soil materials varies widely. Some layers within the O_2 horizon may consist almost entirely of the castings of certain soil arthropods derived from superjacent litter levels; other layers at various levels within the A horizon may contain castings with variable percentages of inorganic and organic content. Not only animals mix and aerate the soil. Plant roots force their way through the soil and may cause significant lateral movement of soil particles. Dead roots are also a source of organic matter for the deeper regions of the soil; roots may extend down to 50 m below the surface in some environments.

Living organisms do more than mix and aerate the soil. One might easily get the idea that soils were in a continual state of depauperization, since gravitational water percolates downward, leaching out nutrients. To be sure, some gravitational water does move upward by evaporation and capillary

Figure 8.16
Nutrient cycling within the soil. Nutrients are released from plants and cycled in part in the upper layers of the soil. Nutrients which leach out into the deep layers can be reincorporated in woody plants with deep-reaching roots.

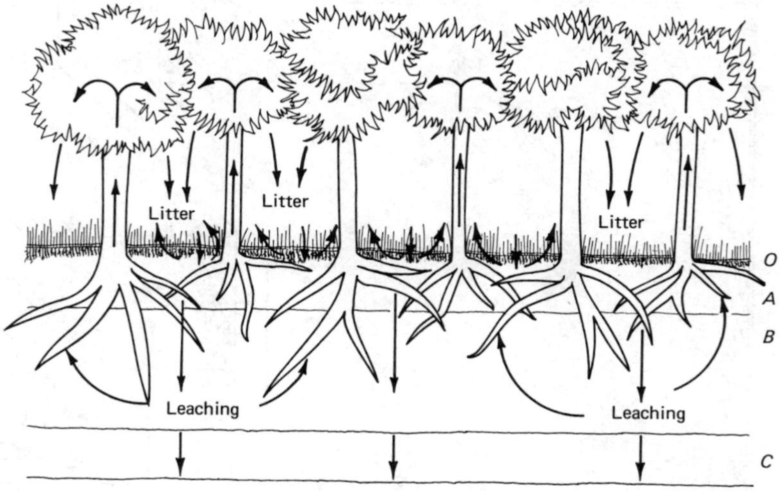

motion in some arid areas, leaching salts to the surface, but this is unusual. For the great majority of ecosystems, if inorganic water movement were the sole means of vertical mixing, the majority of nutrients would be carried down out of the reach of most plants, much as nutrients are lost to the photic zone of the oceans. Deep-rooted plants can utilize nutrients as far down as their roots extend, including the *B* and even the *C* horizons. Nutrients lost to the topsoil can be "recycled" by these plants back into their biomass, thence to the litter layer (Figure 8.16).

Another critical factor in soil formation is erosion, which is a normal process in all terrestrial ecosystems. Erosion has gained a very bad name because human activities have raised erosion rates to such an extent that the long-term stability of intensive agriculture in some of our most fertile lands is questionable.[12] However, erosion at its normal rate (about $\frac{1}{4}$ to $\frac{1}{50}$ mm/yr) is the only means by which mineral nutrients can be made available from their ultimate source, the rock underlying the soil.

Types of Soils

There are many different kinds of soils around the earth, and they influence their ecosystems in many ways. The soils of wet regions are extensively leached, whereas desert soils are not leached at all. The soils of cold regions are both thinner and less iron-rich than those of hot regions. Different soils pose different problems, and the adaptations fixed in populations through natural selection reflect these differences. We shall briefly describe some of the most important soils from different places, pointing out the characteristics that affect terrestrial ecosystems most strongly. The system of classification used is the so-called 7th approximation of the U. S. Department of Agriculture.[13]

There are 10 soil orders, each of which has many subclasses and widespread geographic distribution (Figure 8.17):

1. Immature soils without well developed profile: Entisols
2. Young soils with poorly developed profile: Inceptisols
3. Moderately leached mineral soils of temperate regions: Alfisols
4. Strongly leached mineral soils of temperate regions: Spodosols

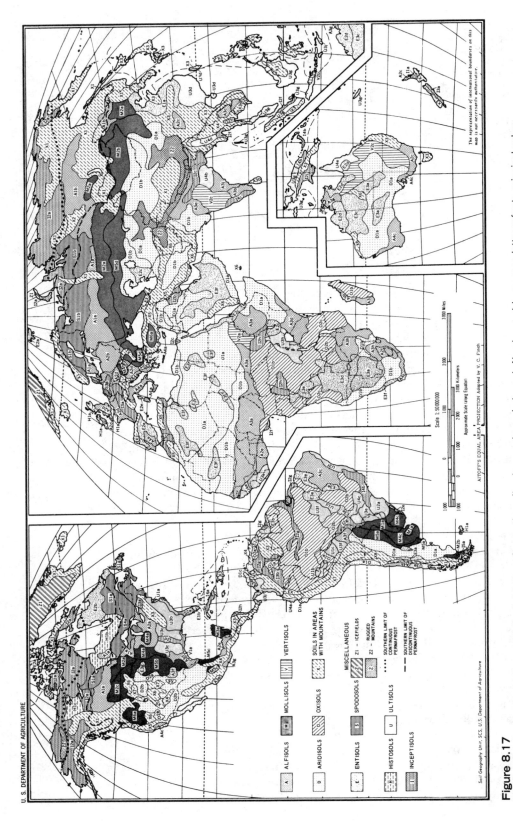

Figure 8.17

Distribution of soils around the world. Capital letter refers to the soil order, as shown in legend. Number and lower-case letter refer to principal suborders. For more information, refer to U.S. Dept. of Agriculture, 1960. *Soil Classification, A Comprehensive System, 7th Approximation.* Washington, D.C.: Soil Conservation Service.

5. Strongly leached neutral mineral soils of warm areas: Ultisols
6. Very strongly leached mineral soils: Oxisols
7. Dark soft soils of high inherent fertility: Mollisols
8. Soils of arid lands and deserts: Aridisols
9. Unstable clay-rich soils: Vertisols
10. Organic soils: Histisols

One of the major factors differentiating soil types is the degree of reworking of the soil from its parent mineral and organic parent materials. Some soils (the so-called *immature* ones) show very little change, whereas others (the *mature* ones) show little similarity to the original components. The characteristics of the immature soils are largely *inherited* from the parents, and those of the mature soils are largely *acquired* through the soil-forming process. At one end of the spectrum, normal soil-forming forces operate very slowly (or not at all) in some environments, so that soils consist of little more than broken-down rock material or undecayed organic detritus; this is the case for soils of deserts and peat bogs. The characteristics of these soils are virtually all inherited from the parent material. At the other end of the spectrum, a soil whose parent materials have been so intensely weathered that any original chemistry of the mineral grains has been obliterated has acquired its characteristics from the environment in which it forms. In young soils, inherited characteristics predominate; conversely, the characteristics of most mature soils are acquired. One can anticipate that any given class of immature soils is more variable than a class of mature soils, and this is, on the whole, true.

Entisols are very young soils that do not have a well-differentiated soil profile. They include a wide spectrum of soils from many different environments. They include the very fertile alluvial soils that made the Nile valley the breadbasket of the ancient world. On the other hand, the soils of very steep hillsides are sometimes little more than broken-up rock; the rate of erosion is so fast that the normal soil-forming processes never get very far. In very dry areas, sand dunes may never be sufficiently vegetated to accumulate any humus.

The most important entisol from a practical point of view is the *alluvial* soil, which consists of sediment carried by rivers and is deposited during periods of flooding. It is rich in colloids—it may even consist primarily of soils eroded from upstream areas—and it may be rich in nutrient materials because it is not leached. Most widespread alluvial soils are found in the flood plains of great rivers and have historically been among the most fertile of soils. Tragically, many areas of alluvial soils, such as the lower Nile and the lower Mississippi, are in areas where soils are easily leached. Efforts at flood control in these areas may have the undesirable result of allowing most of the nutrient materials to become leached out of the soil without being replaced by new materials.

Inceptisols are young soils with poorly developed profiles that form quickly from parent materials. Some of the most productive of these form from volcanic ash and are found in western North America and northern South America. Weathered floodplain soils along the Amazon and Ganges also belong in this class. This group also includes some light colored forest soils of fairly low fertility from eastern North America, western Europe, Siberia, South America, China, north Africa, and India.

Alfisols are moist mineral soils that show better development than the inceptisols but which are less leached than more mature soil types. They are typically formed under deciduous forests, although some have natural grassland vegetation. These include some of our best-known and most productive agricultural soils. They have served Western civilization well, and most industrial countries are built on alfisols, including the northeastern United States, much of Europe, and much of the Asiatic USSR. The fertility of alfisols is moderately high. They are typically gray to brown in color, with the horizons well developed.

Spodosols are characterized by an accumulation of organic matter in the subsoil, along with aluminum and often iron oxides. This horizon commonly occurs just beneath a highly leached, distinctly separate, and typically highly colored leached zone. Spodosols are soils of cool temperate climates, forming under forests on acidic, easily leached parent material. Conifer forests and other vegetation low in nutrient ions encourage the development of spodosols. These soils are highly leached and acid. Most nutrient ions have been removed from the A horizon, leaving it an ashy gray siliceous layer beneath a relatively thick litter layer. The B horizon is typically well developed and thick. They are low in native fertility and are marginal for most crops, except for certain varieties such as potatoes in Maine.

Ultisols are highly weathered mature soils of warm climates. They are more acidic than the alfisols but not as much as the spodosols. They commonly form under forests, although savannas and tropical grasslands are also found on them. Their native fertility is fairly low, although they are easily worked and can be productive if fertilized. They are typically yellow or red, indicating a very high degree of leaching and accumulation of iron oxides.

The most highly weathered soils are the *oxisols*. Unlike the other highly weathered soils (spodosols and ultisols), weathering of the mineral soils has proceeded about as far as it can go, so that they are very high in clay content, and the clay minerals in the soil are iron and aluminum oxides that form under the most intense weathering conditions. These are soils of the wet tropics. It is in the areas with these soils that the detritus food chain is quickly able to break down virtually all of the organic detritus that accumulates in the ecosystem, so that there is virtually no litter layer and very little organic material of any sort in the soil. The A horizon is a very thin brown or gray layer that is easily lost to erosion. The B horizon is typically orange to red, because of the high iron oxide clay content, and it may be extremely thick—some oxisols have a B horizon over 20 m thick.

Vegetation on oxisols tends to be lush tropical rainforests or savannas. It is often assumed from the amount of vegetation that these soils are very rich. In fact, weathering is so complete that they are as nutrient-poor as any soil class found on earth; the lushness of vegetation results from the extremely high rainfall and temperatures and from the exceedingly efficient cycling of what nutrients are available. The adaptations of species in the community compensate very effectively for the limitations of an impoverished soil. Virtually all of the ecosystem's nutrient pool is in the organisms themselves, and they turn it over exceedingly rapidly. One type of oxisol is known popularly as *laterite*. It has the texture of soil when shaded by a forest so that it does not dry out during dry periods, but it hardens irreversibly into a cementlike substance if the forest is cut and the soil dries out. Laterite has been used to construct roads and buildings, including Cambodia's Angkor Wat.

Managing oxisols is not easy. The efficient nutrient recycling mechanisms found in the natural ecosystem do not work when the vegetation is cleared and the soil can leach readily. Certain crops such as rubber, bananas, and cassava can be raised successfully on these soils, but their poverty makes most techniques and crops of Western agriculture inapplicable to them. Unfortunately, many of the crops that grow best on oxisols are protein-poor. One of the chief reasons for the intense poverty of many of the countries in the world's tropical region is the fact that the oxisol soils underlying them cannot sustain a sufficiently high-quality agriculture to allow the areas to feed themselves and build a modern economy at the same time.

Mollisols have a thick, dark, fertile A horizon. They are not weathered nearly as much as the soils discussed so far, and they are found in areas of moderate rainfall under grasslands and prairies. Potential evapotranspiration and rainfall are fairly close, but the available water can maintain a rich flora of mixed tall and short grasses. Productivity is sufficient to produce abundant organic detritus combined with a minimum of leaching, so that the A horizon is unusually thick and rich in organic colloids (up to 15% by weight). Because of the abundance of organic materials with their very high ion exchange capacity and a low level of leaching, the soils are rich in all mineral nutrients, as well as nitrogen and sulfur. Even some very soluble materials such as sodium and potassium may not be uncommon in the lower B horizons of these soils. Calcium is not leached out of the soil, as it is in soils of wetter climates. Rather it collects as a calcium carbonate or calcium sulfate layer in the B horizon.

Mollisols include some of the world's most fertile and important soils. They are easily worked, as well as being fertile. As a result, they are among the most productive agricultural soils in the world, even without chemical fertilizers. They are found in the Great Plains of the United States and Canada, the steppes of the Ukraine and southern Asiatic USSR, and parts of South America. These are the breadbaskets of their respective continents. However, because rainfall is relatively low and unreliable, they may need some irrigation for maximum productivity.

Aridisols are soils of very dry climates. Potential evapotranspiration is much greater than precipitation. They are thin, and a layer of cemented calcium carbonate often lies at or close to the surface of the soil. They are light-colored and may be red in hot deserts. Although there is little leaching in desert and brown soils, the productivity of the areas in which they are developed is so low that detritus production and organic colloid content are also extremely low, and the soils are not very useful for agricultural purposes without extensive irrigation. These soils are found in all of the world's deserts.

Vertisols are characterized by clay minerals that swell markedly, leading to wide, deep cracks in the dry season.

Histisols are the very organic soils found in bogs and related (generally waterlogged) environments. They contain at least 20% organic material, and some histisols contain much more. Areas underlain by histisols are characterized by very low levels of organic decomposition. Decomposition is often so low, in fact, that one can recognize the original form of the fibrous organic matter. These soils represent the storage of considerable amounts of fixed carbon, and they are the source of peat and coal. When drained, they are also very useful for agricultural specialty crops, notably vegetables.

Architecture of Terrestrial Communities

The architecture of a terrestrial community is very different from that of an aquatic community, and it is significantly more complex. This architecture determines not only the community's appearance, but also many of the ways in which different species interact with one another. Several aspects are of importance to us: the *stratification* of the community, and the basic adaptive types, or *growth forms*, of the plants and animals. The stratification of a terrestrial community is determined by its plants. They are rooted to their substrate, and they are typically much larger than aquatic plants. A large proportion of the overall environmental grain is related to the types and distribution of the plants. Indeed, the distribution of animals can often be best understood as occurring within the framework defined by the plants.

Stratification in Terrestrial Communities

A typical forest contains several species of trees, a number of shrubs, and a host of low herbs. The plants belong to different species, and they occupy strikingly different volumes within the forest. If the heights of mature plants in a forest are plotted, they fall into several distinct zones (Figure 8.18). The *canopy* of the forest is its uppermost level. It is exposed to the full insolation from the sun, and is not protected by other forest strata from the buffeting of the winds or rain. It may be a *closed canopy* in which crowns of adjacent trees touch, and direct light does not get through, or it can be an *open canopy* in which substantial amounts of light can pass between adjacent crowns. The *shrub layer* includes the woody plants below 3 m high. Between the canopy and the shrub layer may be one or more strata of smaller *understory* trees. At the ground level is a layer of *herbs* which may extend $\frac{1}{8}$ to $\frac{1}{2}$ m above the surface. Finally, plants such as mosses, liverworts, lichens, and algae, may encrust the soil or exposed rock surfaces.

Different species with different adaptations characterize each level, to

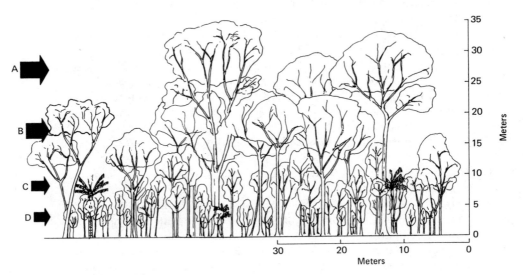

Figure 8.18
Profile diagram of the Crappo-Guatecare climax forest in Trinidad, showing stratification in a complex rainforest. [Redrawn, with permission, from P. W. Richards, *The Tropical Rain Forest: An Ecological Study.* Copyright © 1957 by the Cambridge University Press.]

Figure 8.19
Stratification of various physical variables within a fir forest. [Data from Geiger, 1966.]

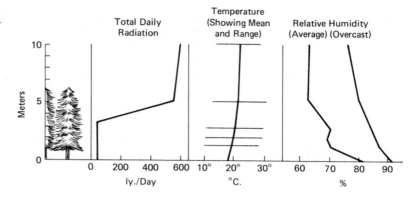

meet the different environmental conditions in each story. Strata differ most prominently with regard to light. Each level absorbs, reflects, reradiates, and diffuses light. The spectrum of light absorption by leaves is quite selective: The wavelengths utilized in photosynthesis are absorbed preferentially over other wavelengths. Thus the spectral quality and intensity of light both change as one goes from the crown to the ground. It is not unusual in some deep forests for the light reaching the forest floor to be a fraction of 1% of the insolation at the forest top, and nocturnal animals such as bats may be about during the day in tropical rainforests.

Air circulation patterns in a forest are also related to community stratification. Plant leaves present a barrier to air movement and tend to retard circulation. The interior of a multistratum tropical rainforest may be virtually windless. Relative humidity is somewhat higher in the lower strata of the forest than in the canopy, and temperature is less variable in the lower stories. Many environmental factors are much less variable at ground level than they are in the canopy. Figure 8.19 shows the average summer values for several climatic variables in a dense fir plantation whose architecture is relatively simple. A more complex forest would show a similar picture, except that each story would contribute to the overall result.

Microhabitat variations within a stratified community place very different selective pressures on the species found in each level. Plants that can occupy the canopy, for instance, have a powerful selective advantage over lower trees with regard to light. They can receive all of the energy they need for photosynthesis, and light is never limiting. But they gain this advantage by tying up a great deal of energy in the woody structure of the trunk, which supports it not only against the force of gravity, but also against the force of the wind. In addition, the higher rate of energy absorption in the canopy means that transpiration is higher; this must be balanced by a more extensive root system to provide water to meet transpiration losses. Clearly, attaining the canopy is not an unalloyed blessing.

Those plants that cannot reach the canopy for one reason or other must either be able to utilize the lower light levels (sometimes much lower) that prevail in lower stories, or sprout and bloom before the trees have expanded their leaves and blocked off the sun, as with spring flowers. In either case, lower-story plants do not need to tie up as much energy in woody tissue, and they live in a somewhat more moderate habitat. Plants that are sensitive to desiccation can exist in the herb layer, protected from the drying effects of high solar radiation and low humidity, if they are sufficiently efficient at photosynthesis to produce enough carbohydrates at attenuated light levels. Some very sensitive plants such as ferns have arborescent representatives in the understory tree strata in warm moist forests which cannot survive if the canopy is removed.

The framework of forest stratification is a useful model for visualizing some of the mechanisms of forest succession, especially in ecosystems that have been cleared by people. The seedlings of all upper-story trees must pass through the lower stories before they reach maturity. The first trees to become established on a field are those that grow rapidly in a sun-drenched environment. The canopy provided by these trees is often open, but even an open canopy reduces the light reaching the ground to a level so low that the seedlings of the same species do not survive. Immature trees with shade-tolerant seedlings may sprout and remain dormant at shrub height until a canopy tree dies. The increased sunlight produced in the clearing by the fall of the tree spurs the growth of all of the saplings in the immediate area, and the original canopy tree is replaced by a different species whose young are more shade-tolerant. This replacement may continue through several cycles. The climax forest consists of upper-story trees whose young tolerate the shades found in the climax. Their young grow slowly in the shaded conditions, but spurt rapidly when a place in the canopy is available and the sunlight in a local area increases.

Table 8.3
Major Growth Forms for Land Plants

Trees: Larger woody plants, mostly well above 3 m. in height, inhabiting canopy and understory layers
  Needle-leaved, mainly conifers
  Broadleaved evergreen, mainly tropical and subtropical forms, most of which have medium-sized leaves
  Evergreen-sclerophyll, with small, tough, evergreen leaves
  Broadleaved deciduous, shedding leaves in tropical dry season or temperate-zone winter
  Thorn-trees, armed with spines, in many cases with compound deciduous leaves
  Rosette trees, unbranched, with a crown of large leaves as in palms and tree ferns
Lianas: Woody climbers or vines
Epiphytes: Plants growing wholly above the ground surface, on other plants
Shrubs: Smaller woody plants, mostly below 3 m. tall
  Broadleaved deciduous
  Evergreen-sclerophyll
  Rosette shrubs
  Stem-succulents, such as the cacti
  Thorn-shrubs
  Semishrubs, with the upper parts of stems and branches dying back in unfavorable seasons
  Subshrubs or dwarf shrubs, spreading near the ground surface, less than 25 cm. high
Herbs: Plants without perennial above-ground woody stems
  Ferns
  Graminioids, the grasses, sedges, and other grasslike plants
  Forbs, herbs other than ferns and graminioids
Thalloid plants: Very low nonvascular plants with an encrusting habit
  Lichens
  Mosses
  Liverworts
  Algae

Modified, with permission, from R. H. Whittaker, *Communities and Ecosystems.* New York: The Macmillan Company, 1970.

Of course, not all ecological succession in forests can be explained by light. But this mechanism has a significant role in many communities, and it can be of considerable practical significance in disciplines such as forest and game management.

Growth Forms of Plants

Growth forms are the functional adaptive types of plants that occur in terrestrial communities. They include features such as position in stratification, type of branching, periodicity (evergreen, semideciduous, deciduous, or leafless) and leaf type (size, outline, and texture). Growth form is a descriptive concept, but it is useful for both unifying observations and making predictions. The most significant growth forms of terrestrial plants are shown in Table 8.3.

It is useful to think of a plant's growth form as the key to its ecological niche. This does not mean that all plants of a given growth form occupy approximately the same niche. But when one examines ecosystems with similar environmental characteristics, the distribution of growth forms is remarkably similar. This underscores the integration of the biotic and abiotic phases of the ecosystem, with the fundamental abiotic factors of climate and geography setting the tone for the community. This allows us to speak of broadly based ecosystem types whose niche structure and functional (but not taxonomic) makeup are rooted in the fundamental abiotic factors of the environment. This allows a much better overview of terrestrial communities than would be the case if each had to discussed separately.

Adaptive Types of Animals

The concept of growth forms in animals is much less well developed than in plants, probably because classification is made more difficult by the fact

Table 8.4
Major Adaptive Types for Land Vertebrates

Herbivores	*Carnivores*
Ground-dwelling	Ground-dwelling
Running	Large cursorial carnivores: wolves, varanid lizards
Large narrow-hooved ungulates of the grasslands	Small cursorial carnivores: weasels, civets, lizards
Large broad-hooved ungulates of moist forests	Large springing carnivores: cats
Large running birds: ostrich, rhea, emu, etc.	Omnivores: raccoons, bears
Jumping	Legless carnivores and insectivores: snakes
Small gnawing animals: rodents, rabbits, wallabies	Large insectivores: anteaters
Large fast-moving animals: kangaroos	Small insectivores: toads, lizards, salamanders
Heavy	Carrion feeders: hyaenas, etc.
Armored slow-moving animals: armadillos, turtles	Digging
Unarmored exclusively terrestrial animals: elephants	Large fossorial carnivores: badgers, etc.
Unarmored amphibious animals: hippopotami	Small fossorial carnivores: ferrets, some snakes
Digging	Aerial
Small gregarious animals: prairie dogs, etc.	Small insectivores: some birds, bats
Tree-dwelling	Carrion feeders: vultures, buzzards, condors
Small gnawing nut-and-leaf eaters: squirrels, etc.	Large raptors: hawks, owls, eagles
Large fruit-and-leaf eaters: monkeys, opossums, etc.	Amphibious
Flying	Large armored carnivores: alligators
Large fruit-eaters: bats	Small insectivores: frogs
Small nut-eaters: birds	
Small nectar-drinkers: hummingbirds	
Small parachuters and gliders: flying squirrels, etc.	
Water and shore birds	

that animals move. But animals have adaptive types analogous to the growth forms of plants. When we compare similar ecosystems, we can see the same striking similarities in the spectrum of animal adaptive types that we see in the growth forms in plants. Adaptive types include features such as place in the food chain, position within the stratification of the community, means of moving, size, and certain behavioral characteristics. Table 8.4 shows the more important adaptive types for vertebrates. Remember that this table deals only with one group of animals. Other groups have a significant role in terrestrial ecosystems. In number of species, the insects are even more important.

Biome Types

The variation in terrestrial ecosystems is infinite, and there are seldom, if ever, clear-cut boundaries between them. But many species typically occur together, and we can speak meaningfully of types of ecosystems, based on the similarity of their composition or structure. The basic ecosystem types are called *biomes*. They tend to be distributed in areas of similar geographic characteristics. They have practical as well as descriptive significance, since they provide a framework within which to interpret differences or similarities between communities. We can also compare a community presently occupying an area that has been altered by human manipulation with the community that would have been present had people not interfered with the natural ecosystem.

There are several schools of thought about the classification of terrestrial community types. Most are based on the distribution of vegetation, both because plants, especially trees, are the most prominent parts of the terrestrial community and because animals often have somewhat broader ranges than plants and hence can cut across different vegetational groupings. Many plant assemblages are found in several localities (Figure 8.20). When this happens, they tend to be associated with the same suite of animals and the same geographic conditions. Conversely, different parts of the world with similar environmental conditions may be inhabited by communities of remarkably similar structure, even though they are completely unlike in species composition. Ecological equivalence (page 134) between members of totally unrelated groups may be prominent.

A detailed comparison of the principles and assumptions underlying the several different classificatory schemes now in use is beyond the scope of this book. They are discussed by Robert Whittaker,[14] and detailed descriptions of the methods of classification are given by numerous other authors.[15]

The basic unit in most systems of classification is the *plant association*, which is determined by observation. The association is assumed to be an assemblage of definite composition and uniform appearance, growing under uniform habitat conditions. However, workers in ecosystem classification cannot agree on standards for permissible variation in species composition, appearance, or habitat. This should not be surprising. When you strip away

Figure 8.20 (opposite)
World distribution of biome types. [Redrawn, with permission, from Pierre Dansereau, *Biogeography—An Ecological Perspective.* Copyright © 1957, by The Ronald Press Company, New York.]

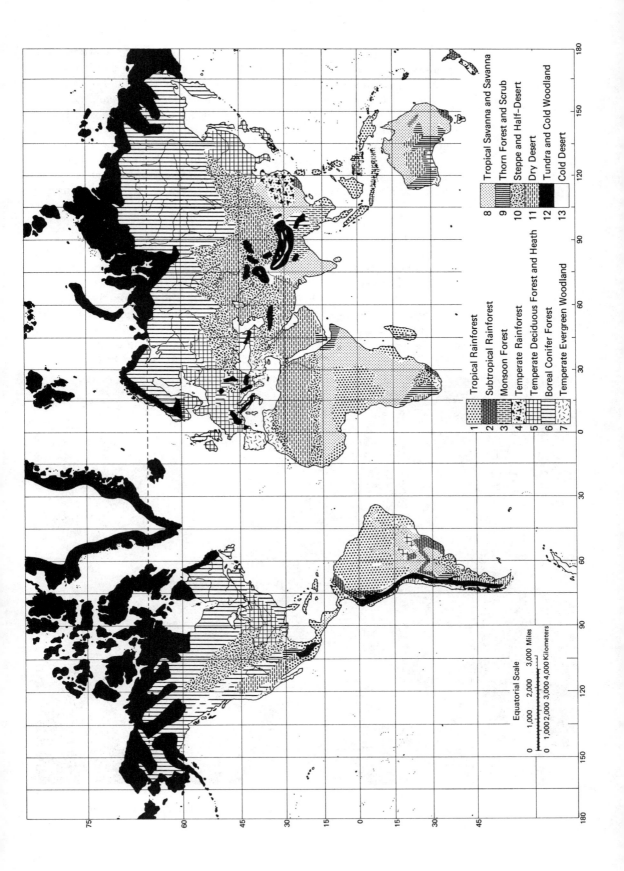

Tropical Savanna and Savanna

Thorn Forest and Scrub

Steppe and Half–Desert

Dry Desert

Tundra and Cold Woodland

Cold Desert

8
9
10
11
12
13

Tropical Rainforest

Subtropical Rainforest

Monsoon Forest

Temperate Rainforest

Temperate Deciduous Forest and Heath

Boreal Conifer Forest

Temperate Evergreen Woodland

1
2
3
4
5
6
7

Equatorial Scale

0   1,000   2,000   3,000 Miles

0 1,000 2,000 3,000 4,000 Kilometers

the jargon, the purpose of ecosystem classification is to create an order out of the myriad of signals flowing back and forth in ecosystems and whose net results are manifested most prominently in the community. For our purposes here, it is significant that the complexity of ecosystems has overwhelmed the classificatory bent of many biologists around the world. It is also meaningful for us to describe communities in the less precise notion of biomes, which are generalized types of communities comprising several associations with similar community structure. They include animals as well as plants. Thus, the deciduous broadleaf forest communities of eastern North America constitute a biome, although it may include several recognizable associations.

Tropical Rainforest

Tropical rainforests are among the most diverse communities on earth. Both temperature and humidity are high and constant. The trees are tall and often have buttressed bases. There are typically two understory tree levels, but the shrub and herb layers are not particularly well developed. Epiphytes and vines are common. Tree diversity is extraordinarily high: A hectare seldom contains less than 20 different species of tall trees, and may contain over 50. Nearly all plants are evergreen, and those that do lose their leaves entirely do so at irregular intervals with no apparent regard to the climatic

Figure 8.21
Photograph of a tropical rainforest. Loquillo Experimental forest, Puerto Rico. [Photo courtesy of the United States Forest Service.]

regime. The leaves of most plants are of moderate size, leathery, and dark green in color. Soils are red oxisols and may be exceedingly thick. The high rate of leaching makes these soils virtually useless for agricultural purposes, but if they are left undisturbed, the extremely rapid cycling of nutrients within the litter layer can compensate for the natural poverty of the soil.

Invertebrate density and abundance are very high in tropical rainforests. Although vertebrates are diverse, they are not as abundant as in many other communities. They also tend to be somewhat smaller, and large animals of nearby grasslands may have pigmy relatives in rainforests. The arboreal habitat seems also to be characteristic of rainforests. Where they exist, monkeys are abundant in the natural fauna. In New Guinea and northern Queensland, where monkeys are absent, there are arboreal kangaroos (*Dendrolagus* sp.), despite the fact that the basic anatomy of the kangaroo is not particularly well suited to arboreal life.

Subtropical Rainforests

Subtropical rainforests are found in humid subtropical areas characterized by a noticeable, but not pronounced, difference between summer and winter. Seasonality is of a wet-dry type rather than a hot-cold type. Stratification is simpler, with only one understory tree horizon. Plants tend to be evergreen but may lose their leaves during the dry season. Epiphytes and vines are noticeably scarcer than in tropical rainforests, but the animal life is very similar.

Tropical Seasonal Forests

Tropical seasonal forests occur in regions whose total annual rainfall is very high but segregated into pronounced wet and dry periods. Exceedingly wet tropical seasonal forests are popularly known as monsoon forests; their annual precipitation may be several times that of the tropical rainforest. Trees may reach heights over 40 m but are more commonly 20 to 30 m high. Stratification is relatively simple, with a single understory tree layer, and there is often a pronounced difference in growth form between these strata, with the canopy being deciduous and the understory being evergreen. Teak is a major large tree in the best-known tropical seasonal forests, those of India and Southeast Asia. However, its natural role in forests is quite uncertain. Its value as a timber tree may have led people to introduce it into areas in which it did not occur naturally. Regardless of its origin, it is now a dominant tree in many Southeast Asian tropical seasonal forests. Bamboo is also an important climax shrub in these areas, although in other areas it is important only in earlier stages of the succession. Tropical seasonal forests are also found in Central and South America, northern Australia, western Africa, and the tropical islands of the Pacific, as well as India and Southeast Asia.

Temperate Rainforests

The temperate rainforest is the coldest of the rainforests. Its climate is unquestionably seasonal, with temperature and rainfall both varying throughout the year. Rainfall is high, but fog is often heavy and can actually be a more important source of water for living organisms than rainfall. Diversity of both plants and animals is lower than in other rainforests, but it is considerably higher than in other temperate forest types. The canopies in-

Figure 8.22
Photograph of a
temperate rain-
forest, Olympic Na-
tional Forest, Wash-
ington. [Photo
courtesy of United
States National
Park Service.]

clude the tallest trees in the world, such as the coast redwood (*Sequoia sempervirens*) of the Pacific coast of North America and the alpine ash (*Eucalyptus regnans*) of Australia and Tasmania, both of which reach more than 100 m in height. Epiphytes and vines are common, but again, less abundant than in warmer rainforests. The animals of temperate rainforests are similar to those of temperate deciduous forests but show a somewhat higher diversity.

Temperate Deciduous Forests

Moist temperate seasonal climates are characterized by deciduous forests. Seasonality is hot-cold, and precipitation may either be seasonal or fairly uniform throughout the year. Trees are quite tall—about 40 to 50 m in height—and their leaves are thin and broad. Soils are alfisols or spodosols, and they are fairly deep. There are few epiphytes save for some species of mosses, algae, and lichens growing on tree trunks. Vines are also uncommon except for a few varieties such as *Vitis*, the grape.

Figure 8.23
Photograph of a
temperate decidu-
ous forest. Mixed
hardwood forest in
the Shawnee Na-
tional Forest, Illi-
nois. [Photo cour-
tesy of United
States Forest Ser-
vice.]

This biome type includes the northeastern United States and most of western Europe. The canopy is moderately diverse. It is dominated by two or three species, but it contains several others as well. A patchy to well-developed understory tree layer is often found, and shrubs are common and represented by a large number of species. There are often two separate herb assemblages. One consists of spring flowers which bloom before the trees have expanded their leaves and are gone by summer. The other is adapted to the low light levels of the forest floor and lasts into the fall.

Animals as well as plants are seasonal; some may even hibernate throughout the winter. Animals range widely in size and adaptations. The largest include deer and bear. Large carnivores include animals such as the wolf and mountain lion. There are also numerous smaller carnivores and omnivores such as fox, skunk, and raccoon. Dominance by a few species is characteristic of the temperate deciduous forest, and diversity is lower than in any of the rainforest types.

Boreal Conifer Forests

Cold regions with high rainfall and strongly seasonal climates with long winters and short summers are characterized by boreal conifer forests. Many animals and plants are large: Trees range up to 40 m in height, and the vertebrates include the giants of several groups, such as moose, caribou, reindeer, elk, grizzly bear, wolverine, beaver, and several large species of birds. Species diversity is low, and pure stands of trees and shrubs are com-

Figure 8.24
Photograph of a
boreal conifer for-
est. A mature
spruce-fir stand in
the Lolo National
Forest, Idaho.
[Photo courtesy of
United States For-
est Service.]

mon, as are natural outbreaks of defoliating insects. Understory trees are uncommon, but the shrub layer is often quite rich, as are the thalloid mosses and lichens. Soils are thin spodosols, and their fertility is poor, both because rock weathering proceeds slowly in the cold environments and because the litter derived from conifer needles breaks down very slowly and is not partic-ularly rich in nutrients. The productivity of a boreal conifer forest is the lowest of any forest biome.

Temperate Evergreen Woodland

Many parts of the world have a Mediterranean-type climate with warm, dry summers and cool, moist winters. These are characterized by low ever-green trees with small, hard needles or slightly broader leaves. The best known areas of tropical evergreen woodland are the chaparral of the Pacific coast in North America, the Mediterranean *maquis*, and the Spanish *encinar*. Trees are essentially lacking, although shrubs may range up to 3 to 4 m in height. Species diversity is roughly intermediate between that of a temperate deciduous forest and a drier grassland.

Figure 8.25
Photograph of a
temperate evergreen
woodland. Chaparral
stand in the Cleve-
land National For-
est, California. Mt.
Palomar is in the
background. [Photo
courtesy of the
United States For-
est Service.]

Fire is an important factor in this ecosystem, and many plants are closely adapted to the fire-prone environment. Not only do they regenerate quickly after being burned, many of them are so flammable that areas occupied by them burn readily and often. As a result, fire is so common that only plants that regenerate quickly after fires can compete successfully. The woodlands are much more open than true forests, and visibility is much greater. This puts a premium on speed of motion for animals, whether they be herbivores or carnivores. Small-hooved cursorial ungulates are the dominant herbivores, and some are quite large. Jumping animals also constitute a significant portion of the fauna.

Tropical Savanna

The savanna is intermediate in many ways between a tropical forest and a grassland. Rainfall is very sporadic. The biome type is extremely variable and is often subdivided. Trees are present and may range up to 20 m in height, but they are commonly closer to 10 m high. They are sparse and come nowhere near forming a canopy. Savanna trees are resistant to desiccation, and may be either deciduous or evergreen. Their leaves are often hard and drought-resistant. Grasses are the most conspicuous plants and may reach heights of 1½ to 2 m. The largest and most typical area of tropical savanna is in east Africa, where it supports the greatest diversity of grazing vertebrate life to be found anywhere in the world. But savannas occur in Australia and South America as well as Africa.

Gigantism of certain groups in these tropical savannas is as pronounced as it is in the boreal forest, including such giants as many antelopes, giraffes, elephants, buffalo, and lions. Savanna soils are ultisols and oxisols.

The distribution of savannas shows more clearly than most the interplay between the fundamental abiotic environmental variables and the other signals operating in the ecosystem. Savannas are found in warm, dry climates,

TERRESTRIAL
ECOSYSTEMS

249

Figure 8.26
Photograph of a
tropical savanna.
Grassland in the
Kruger National
Park, South Africa.
[Photo courtesy
M.B. Clapham.]

and they are typical of those climates. but more than climate is involved in their distribution. One can generally find a forested area with the same climatic or geographic parameters as a savanna. Fire is a common occurrence in savannas, and it seems likely that the savanna as a biome is maintained by fire (perhaps aided by intense herbivore pressure), which gives grasses and certain species of trees a powerful advantage over other tree species.

Figure 8.27
Photograph of a
thorn scrubland.
Tall dense sage-
brush stand in Ruby
Valley, Nevada.
[Photo courtesy of
United States For-
est Service.]

Thorn Woodland and Scrubland

The climate of a thorn woodland or scrubland is strongly seasonal, with the dry season much better marked than the wet. The communities are exceedingly variable; woodland, in which trees may exist up to 5 to 10 m in height, merges with scrubland, whose shrubs average $\frac{1}{2}$ to 1 m high, at frequent intervals. The dominant plants tend to be armed with thorns and are often

succulents such as cacti. All plants have well-developed means of reducing water losses. Surprisingly, a few epiphytes, including Spanish moss (*Tillandsia recurvata*) and some orchids (*Jonopsis* sp.) may be present, but grasses are very uncommon. The entire herbaceous layer, in fact, may be poorly developed. Species diversity of both plants and animals is very low. Soils are poorly developed aridisols, quite dry through exposure to the sun, and calcified.

Temperate Grassland

Grasslands show pronounced climatic seasonality, with respect both to rainfall, which is concentrated in the summer, and to temperature. They include the prairies, great plains, and arid grassland of North America as well as the steppes of Eurasia, the veldt of Africa, and the pampas of South America. Stratification is reduced essentially to a single story, but within that level, species diversity may be as high as in a deciduous forest, especially for the tallgrass prairies. Only along streams are trees to be found, but the "gallery forests" within a few meters of the stream bank are a characteristic of grasslands.

The grasses composing the grasslands can be divided into two basic groups: the tall grasses over 1 m high, which are found in moister portions of the grassland, and the short grasses less than 1 m high, which are found in the drier regions. The soils of the grasslands are rich, fertile mollisols. Typical animals tend to be quite small, except for a few kinds of large running herbivores such as the bison and pronghorn in North America, the wild horse, ass, and Saiga antelope of Eurasia, and some of the antelopes of southern Africa. The large herbivores are nowhere near as diverse as they are in savanna areas. Likewise, the carnivores are relatively small, such as coyotes, weasels, badgers, foxes, ferrets, owls, and rattlesnakes. Rodents, especially

Figure 8.28
Photograph of virgin prairie grassland. This photograph was taken in 1870 at Casper Mountain in southeastern Wyoming. The dominant vegetation today is sagebrush. [Photo courtesy of United States Soil Conservation Service.]

Figure 8.29
Photograph of a
desert in the Tonto
National Forest, Ari-
zona. [Photo cour-
tesy United States
Forest Service.]

burrowing colonial rodents such as prairie dogs, are common. Jumping motion is widespread both in mammals (e.g., rabbits and kangaroo rats) and in
insects (e.g., grasshoppers and crickets). Jumping increases the rate at which
the animal can move through (or over) the tall grasses and makes it harder for
a predator to catch the animal. Visibility in the grasslands is exceedingly
good, and the premium placed on efficiency in eluding predators is very high.

Deserts

Deserts are the biome formed in the driest of environments. Temperatures range from very hot, as in subtropical deserts, to very cold. Our typical
picture of desert is the hot deserts such as the Sahara, the Namib in southern
Africa, or the deserts of the southwestern United States and northern Mexico. Cold deserts include the Great Basin of North America and the Gobi of
Asia. Vegetation is sparse. What plants exist tend to be shrubs with well-
developed adaptations for conserving water, and rapidly growing annuals
which can run through their entire life cycle on the water delivered by one
rainy spell. Deserts may be virtually lifeless, as in the very dry subtropical
deserts, where a few plants that can survive the shifting sands send roots
down extremely deep into the ground. Warm-temperate deserts and semi-
deserts may contain several species of plants, as may cool-temperate scrub
deserts, such as the sagebrush scrub of western North America, and the cold
deserts of the Arctic.

Large animals are very uncommon, although the mule deer and some
species of gazelle may be found in small numbers in scrub areas. Small
rodents are the most common mammals, along with the coyote and small
foxes, but most of the herbivores are insects. Because of this, the number of
small insectivorous lizards is higher in the desert than in other biomes. As in
other open areas, running, digging, and jumping adaptations are widespread
among all animals.

Tundra

Very cold terrestrial ecosystems are characterized by water permanently
frozen in the subsoil. This is *permafrost*. The communities developed in

these areas are adapted to a very short growing season and exceedingly rigorous conditions of cold. Trees are entirely lacking except near rivers, where stunted gallery forests may be developed, and on sheltered slopes, where the unfrozen soil may be unusually deep and scrub and heath can be found. The herb layer is widespread in the tundra and may consist of many different species of plants, particularly grasses and sedges. The lichen layer covering exposed rocks is better developed than in any other biome type.

Bogs and muskeg are common in the moist areas, because drainage through the frozen soil is universally bad. Large animals are conspicuous, especially the wolf, musk ox, reindeer, barren ground caribou, and grizzly and Kodiak bears. In addition, varying hares, lemmings, owls, and lynxes are common. Waterfowl are seasonally abundant. Diversity is very low, and despite the extreme hardiness of the tundra organisms, their growth rates tend to be very low, and the community as a structural unit is exceedingly vulnerable. It is for good reason that the tundra is the most common example of a "fragile ecosystem."

Figure 8.30
Photograph of a tundra. Slopes of Mt. McKinley, Mt. McKinley National Park, Alaska. [Photo courtesy of United States National Park Service.]

Notes

[1] Mather, J. R., and Yoshioka, G. A., 1968. The role of climate in the distribution of vegetation. *Ann. Assn. Am. Geog.*, **58**, 29–41.

² Carter, D. B., and Mather, J. R., 1966. Climatic classification for environmental biology. *C. W. Thornthwaite Associates Pub. in Climatology*, **19**(4), 305–395.

³ Stone, E. C., and Vasey, R. B., 1968. Preservation of coast redwoods on alluvial flats. *Science*, **159**, 157–161.

⁴ Wright, H. E., Jr., 1974. Landscape development, forest fires, and wilderness management. *Science*, **186**, 487–495; Stone, E. C., and Vasey, R. B., 1968. Preservation of coast redwoods on alluvial flats. *Science*, **159**, 157–161.

⁵ Dodge, M., 1972. Forest fuel accumulation—a growing problem. *Science*, **177**, 139–142.

⁶ McColl, J. G., and Grigal, D. F., 1975. Forest fire: Effects on phosphorus movement to lakes. *Science*, **188**, 1109–1111.

⁷ Wagle, R. F., and Kitchen, J. H., Jr., 1972. Influence of fire on soil nutrients in a ponderosa pine type. *Ecology*, **53**, 118–125.

⁸ Kilgore, B. M., and Taylor, D., 1979. Fire history of a sequoia-mixed conifer forest. *Ecology*, **60**, 129–142.

⁹ Kilgore, B. M., 1972. Fire's role in a sequoia forest. *Naturalist*, **23**, 26–37; Kilgore, B. M., 1973. The ecological role of fire in sierra conifer forests: Its application to national park management. *Jour. Quat. Research*, **3**, 496–513.

¹⁰ Ahlgren, I. F., and Ahlgren, C. E., 1960. Ecological effects of forest fires. *Bot. Rev.*, **26**, 483–533.

¹¹ μ = microns. 1μ = 1/1000 mm.

¹² U.S. Council on Environmental Quality, 1981. *National Agricultural Lands Study: Final Report*. Washington: Government Printing Office.

¹³ U.S. Dept. of Agriculture, Soil Survey Staff, 1960. *Soil Classification, A Comprehensive System*, 7th approximation. Washington, D.C.: U.S. Dept. of Agriculture, Soil Conservation Service. Supplement published in 1967; Brady, N. C., 1974. *The Nature and Properties of Soils*, 8th ed. New York: Macmillan Publishing Company.

¹⁴ Whittaker, R. H., 1962. Classification of natural communities. *Bot. Rev.*, **28**, 1–239.

¹⁵ e.g. Daubenmire, R., 1968. *Plant Communities: A Textbook in Synecology*. New York: Harper & Row, Publishers; Kuchler, A. W., 1964. Potential natural vegetation of the coterminous United States. Am. Geog. Soc. Spec. Pub. 36; Shelford, V. E., 1963. *The Ecology of North America*. Urbana: The University of Illinois Press; Dansereau, P., 1957. *Biogeography: An Ecological Perspective*. New York: The Ronald Press Company; Oosting, H. J., 1956. *The Study of Plant Communities: An Introduction to Plant Ecology*, 2nd ed. San Francisco: W. H. Freeman and Company; Becking, R. W., 1957. The Zurich-Montpellier school of phytosociology. *Bot. Rev.*, **23**, 411–488; Pielou, E. C , 1969. *An Introduction to Mathematical Ecology*. New York: John Wiley & Sons, Inc., chapters 19, 20.

Epilogue

Part III of this book pointed out how understanding a community requires simultaneous juggling of interspecific and intraspecific control mechanism, abiotic factors, and the ecosystem's resource base. That understanding is an abstraction that does well to be put into the real world. Part IV summarizes the basic geographic patterns of natural ecosystems. These patterns are established by the natural landscape, combined with the ecosystem's resource base and the physical characteristics of key materials such as water. Even a brief designation of an ecosystem like "tropical rainforest" or "rocky intertidal" or "profundal benthos" can conjure up an image of the community, even though none of the designations formally indicate the community as such. To be sure, this image does not specify individual species: It suggests life zones and life forms. But this is adequate for many purposes.

Even so, there is a great difference between understanding a biome in a general sense and understanding the operation of a specific ecosystem. Using notions like life zones is a good means of getting oriented to a system, but understanding the system in detail requires greater depth of understanding. What species constitute the community? What are the population characteristics of each species, at least the important ones? What is the complex of signals that binds the community together? Specifically, what spectrum of mutualistic, competitive, predator-prey, or other interactions exist among populations? What role does each population have in nutrient cycling? How does each population influence the overall complex of abiotic variables, and how does it respond to that complex?

Answering these questions means taking very detailed measurements of nutrient levels, population parameters, energy flows, and so on. It means noting the qualitative aspects of the ecosystem in substantial detail. The qualitative image conjured up by a community's life-zone can provide a first-cut approximation of how the community works, but detailed understanding must be quantified. Setting a firm underpinning on our knowledge and understanding of ecosystems was one of the main purposes of the International Biological Programme (IBP). This was an international program designed to study representatives of each of the important biomes of the world and to develop a detailed understanding of their behavior. The IBP was the first large-scale ecosystem study coordinated on a global scale and designed to assist scientists in countries throughout the world by providing a comparative base for ecosystem analysis.

The IBP projects attempted to measure as many signals as possible within the ecosystem and relate them in a systematic way to each other. In general, this was done by computer-based models designed to simulate the behavior of various parts of the ecosystem, given the mix of signals that impinged on it. This behavior included not only the direct responses of the element of the ecosystem in question but also the signals it generated, in turn, to other parts of the ecosystem. The IBP has added immeasurably to our understanding of natural ecosystems. It has also emphasized the degree to which the behavior of living organisms is a response to the abiotic environment, and vice versa. You cannot meaningfully speak of the dynamics of a forest without describing the interaction between the plants and the soils and how that interaction affects nutrient cycling. You cannot speak of soils without describing the vegetation. You cannot make sense out of an aquatic community without first specifying the temperature, oxygen concentration, rate of flow, and turbidity of the water.

Natural ecosystems are incredibly complex. A given species responds to so many different factors that it is often difficult to tell just what is important and what is not. But natural ecosystems are more than a scientific curiosity. The responses of populations to the complex of signals generated by other populations and by their physical geography are more than an arcane study of natural history. Humans, too, are biological populations that require food, water, oxygen, and other essentials of life. The principles that govern natural ecosystems also govern human ecosystems.

Further Reading

Allee, W. C., Park, O., Emerson, A. E., Park, T., and Schmidt, K. P., 1949. *Principles of Animal Ecology*. Philadelphia: W. B. Saunders Co.

Bailey, H. P., 1960. A method of determining the warmth and temperateness of climate. *Geografiska Annaler*, **42**, 1–16.

Bailey, H. P., 1964. Toward a unified concept of the temperate climate. *Geog. Rev.*, **54**, 516–545.

Billings, W. D., and Mooney, H. A., 1968. The ecology of arctic and alpine plants. *Biol. Rev.*, **43**, 481–529.

Blair, W. F., 1977. *Big Biology: The US/IBP*. New York: Academic Press.

Blumenstock, D. I., and Thornthwaite, C. W., 1941. Climate and the world pattern, in U.S. Department of Agriculture Yearbook for 1941, *Climate and Man*, 98–127.

Brown, J., Miller, P. C., Tieszen, L. L., and Bunnell, F. L., 1981. *An Arctic Ecosystem*. Stroudsburg, Penna.: Hutchinson Ross Publishing Co.

Bruun, A., 1957. Deep sea and abyssal depths. Chap. 22 in Hedgpeth, J. W., ed., *Treatise on Marine Ecology and Paleoecology*, vol. 1: Ecology. Geological Society of America Memoir 67, pt. 1, 641–672.

Burges, A., and Raw, F., 1967. *Soil Biology*. London: Academic Press Ltd.

Chapman, V. J., 1977. *Wet Coastal Ecosystems*. Amsterdam: Elsevier Scientific Publishing Co.

Clark, J. R., and Benforado, J., 1981. *Wetlands of Bottomland Hardwood Forests*. Amsterdam: Elsevier Scientific Publishing Co.

Clarke, G. L., 1965. *Elements of Ecology*, rev. printing. New York: John Wiley & Sons, Inc.

Cooper, C. F., 1961. The ecology of fire. *Scientific American*, **204**, 150–160.

Cummings, K. W., 1974. Structure and function of stream ecosystems. *BioScience*, **24**, 631–641.

Daubenmire, R., 1968. Ecology of fire in grasslands. *Adv. Ecol. Research*, **5**, 209–266.

Dietrich, G., 1963. *General Oceanography: An Introduction*. New York: Interscience Publishers.

di Castri, F., Goodall, D. W., and Specht, R. L., eds., 1981. *Mediterranean Type Shrublands*. Amsterdam: Elsevier Scientific Publishing Co.

Doty, M. S., 1957. Rocky intertidal surfaces. Chap. 18 in Hedgpeth, J. W., ed., *Treatise on Marine Ecology and Paleoecology*, vol. 1: Ecology. Geological Society of America Memoir 67, pt. 1, 535–586.

Drew, J. V., ed., 1967. *Selected Papers in Soil Formation and Classification*. Madison, Wisc.: Soil Science Society of America, Special Publication 1.

Edmonds, R. L., 1981. *Analysis of Coniferous Forest Ecosystems in the Western United States*. Stroudsburg, Penna.: Hutchinson Ross Pub. Co.

Edmondson, W. T., ed., 1959. *Fresh-water Biology*, by H. B. Ward and G. C. Whipple, 2nd ed. New York: John Wiley & Sons, Inc.

Emery, K. O., and Stevenson, R. E., 1957. Estuaries and lagoons. Chap. 23 in Hedgpeth, J. W., ed., *Treatise on Marine Ecology and Paleoecology*, vol. 1: Ecology. Geological Society of America Memoir 67, pt. 1, 673–750.

Engelstad, O. P., ed., 1970. Nutrient Mobility in Soils: Accumulation and Losses. Madison, Wisc.: Soil Science Society of America Special Publication 4.

Eyre, S. R., 1968. *Vegetation and Soils: A World Picture*, 2nd ed. Chicago: Aldine Publishing Company.

Geiger, R., 1965. *The Climate Near the Ground*, 4th ed. Cambridge, Mass.: Harvard University Press.

Golley, F. B., and Medina, E., 1974. *Tropical Ecological Systems: Trends in Terrestrial and Aquatic Research*. New York: Springer-Verlag.

Guidry, N. P., 1964. A Graphic Summary of World Agriculture, rev. ed. U.S. Dept. of Agriculture Miscellaneous Publication, **705**, 63 pp.

Halle, F., Oldeman, R. A. A., and Tomlinson, P. B., 1977. *Tropical Trees and Forests: An Architectural Analysis*. New York: Springer-Verlag.

Hesse, R., Allee, W. C., and Schmidt, K. P., 1951. *Ecological Animal Geography*, 2nd ed. New York: John Wiley & Sons, Inc.

Hobson, E. S., 1975. Feeding patterns among tropical reef fishes. *Am. Sci.*, **63**, 382–392.

Holdridge, L. R., 1947. Determination of world plant formations from simple climatic data. *Science*, **105**, 367–368.

Hutchinson, G. E., 1957. *A Treatise on Limnology*, vol. I: Geography, Physics, and Chemistry. New York: John Wiley & Sons, Inc.

Hutchinson, G. E., 1967. *A Treatise on Limnology*, vol. II: Introduction to Lake Biology and the Limnoplankton. New York: John Wiley & Sons, Inc.

Hutchinson, G. E., 1973. Eutrophication. *Am. Sci.*, **61**, 269–279.

Hynes, H. B. N., 1966. *The Biology of Polluted Waters*. Liverpool: University of Liverpool Press.

Hynes, H. B. N., 1970. *The Ecology of Running Waters*. Toronto: University of Toronto Press.

Jenny, H., 1980. *The Soil Resource: Origin and Behavior*. New York: Springer-Verlag.

Jordan, C. F., ed., 1981. *Tropical Ecology*. Stroudsburg, Penna.: Hutchinson Ross Publishing Co.

Jordan, C. F., and Herrera, R., 1981. Tropical rain forests: Are nutrients really critical? *Am. Nat.*, **117**, 167–180.

Kendeigh, S. C., 1961. *Animal Ecology*. Englewood Cliffs, N. J.: Prentice-Hall, Inc.

Kozlowski, T. T., and Ahlgren, C. E., 1974. *Fire and ecosystems*. New York: Academic Press.

Larsen, J. A., 1980. *The Boreal Ecosystem*. New York: Academic Press.

Lauff, G. H., ed., 1967. *Estuaries*. American Association for the Advancement of Science, Publication 83.

Lerman, A., ed., 1978. *Lakes: Chemistry, Geology, Physics*. New York: Springer-Verlag.

Likens, G. E., ed., 1972. *Nutrients and Eutrophication: The Limiting-Nutrient Controversy*. American Society of Limnology and Oceanography, Special Symposia I.

Lindeman, R. L., 1942. The trophic-dynamic aspect of ecology. *Ecology*, **23**, 399–418.

Mangelsdorf, P. C., 1967. Salinity measurements in estuaries, in Lauff, G. H., ed., *Estuaries*. American Association for the Advancement of Science, Publication 83, pp. 71–79.

McIntosh, R. P., ed., 1978. *Phytosociology*. Stroudsburg, Penna.: Dowden, Hutchinson & Ross.

Mutch, R. W., 1970. Wildland fires and ecosystems—a hypothesis. *Ecology*, **51**, 1046–1051.

National Academy of Sciences, 1969. *Eutrophication: Causes, Consequences, Correctives*. Washington, D.C.: National Academy of Sciences.

Odum, H. T., and Pigeon, R. F., 1970. *A Tropical Rain Forest: A Study of Irradiation and Ecology at El Verde, Puerto Rico*. Oak Ridge, Tenn.: Division of Technical Information, U.S. Atomic Energy Commission.

Paton, T. R., and Williams, M. A. J., 1972. The concept of laterite. *Ann. Assn. Am. Geog.*, **62**, 42–56.

Pickard, G. L., 1963. *Descriptive Physical Oceanography: An Introduction*. Oxford: Pergamon Press Ltd.

Pomeroy, L. R., 1974. The ocean's food web, a changing paradigm. *BioScience*, **24**, 499–504.

Pomeroy, L. R., and Wiegert, R. G., eds., 1981. *The Ecology of a Salt Marsh*. New York: Springer-Verlag.

Price, L. W., 1972. The periglacial environment, permafrost, and man. Association of American Geographers Commission on College Geography, Resource Paper No. 14.

Raymont, J. E. G., 1963. *Plankton and Productivity in the Oceans*. Oxford: Pergamon Press Ltd.

Reichle, D. E., ed., 1973. *Analysis of Temperate Forest Ecosystems*. New York: Springer-Verlag.

Reimers, N., and Combs, D., 1956. Method of evaluating temperatures in lakes with description of thermal characteristics of Convict Lake, California. U.S. Fish and Wildlife Service, *Fish. Bull.*, **56(105)**, 535–553.

Richards, P. W., 1957. *The Tropical Rain Forest: An Ecological Study*. New York: Cambridge University Press.

Risser, P. G., Birney, E. C., Blocker, H. D., May, S. W., Parton, W. J., and Wiens, J. A., 1981. *The True Prairie Ecosystem*. Stroudsburg, Penna.: Hutchinson Ross Publishing Co.

Rumney, G. R., 1968. *Climatology and the World's Climates*. New York: Macmillan Publishing Company.

Sinclair, A. R., and Norton-Griffiths, M., eds., 1980. *Serengeti: Dynamics of an Ecosystem*. Chicago: University of Chicago Press.

Steele, J. H., 1974. *The Structure of Marine Ecosystems*. Cambridge, Mass.: Harvard University Press.

Sverdrup, H. V., Johnson, M. W., and Fleming, R. H., 1942. *Oceans: Their Physics, Chemistry, and General Biology*. Englewood Cliffs, N.J.: Prentice-Hall, Inc.

Tieszen, L. L., ed., 1978. *Vegetation and Production Ecology of an Alaskan Arctic Tundra*. New York: Springer-Verlag.

Walter, H., 1979. *Vegetation of the Earth*. New York: Springer-Verlag.

Wells, J. W., 1957. Coral reefs. Chap. 20 in Hedgpeth, J. W., ed., *Treatise on Marine Ecology and Paleoecology*, vol. 1: Ecology. Geological Society of America Memoir 67, pt. 1, 609–632.

Weyl, P. K., 1970. *Oceanography: An Introduction to the Marine Environment*. New York: John Wiley & Sons, Inc.

Whittaker, R. H., 1975. *Communities and Ecosystems*, 2nd ed. New York: Macmillan Publishing Company.

Wimpenny, R. S., 1966. *The Plankton of the Sea*. New York: American Elsevier Publishing Company, Inc.

Prospects

V

THIS BOOK has presented natural ecosystems as complex systems comprising living and nonliving parts. Each part of the system has a role, and the structure of the ecosystem reflects the adaptations of its component populations and the physical and chemical properties of the abiotic environmental factors. The diversity of natural ecosystems is tremendous. Still, we can classify environments into basic types which reflect similar fundamental environmental conditions and which show similar responses to changes in their governing variables. If this were the consummation of the study of natural ecosystems, it would be a significant science. But the study of natural ecosystems goes far beyond this. We are learning enough about the principles that govern ecosystem behavior that we can predict their responses to unusual perturbations and the paths they will follow when perturbed. This is of great practical significance, since people are the most important agent of environmental perturbation at the present time, and many environmental perturbations can affect the biological or economic survival of human society.

Implications
of Natural
Ecosystems

9

WE COMMONLY EXPRESS the responses of an ecosystem to perturbations in terms of its stability. As pointed out in Chapter 6, stability means different things to different people, and the practical implications of different interpretations are quite significant. It is probably most meaningful to use the different meanings of stability pointed out by Gary Harrison.[1] These are constancy (degree of variation around an equilibrium), resistance (inertia to change in time of stress), resilience (tendency to return to the equilibrium when perturbed away from it), and persistence (tendency to remain viable as a system for a long period of time in spite of stress). These notions can be defined rigorously, and all represent dispassionate ways of describing stability.

One can also describe an ecosystem's responses to stress in explicitly value terms. One way to do this is in terms of its "health." This is a very subjective measure that refers to the ecosystem's resemblance to the appearance or structure that would be expected for the area. It is, if you will, a way of expressing a value judgment on the status of an ecosystem. A climax forest might be perceived as a healthy ecosystem; an agricultural field with a simpler community and a lower gross productivity, but whose soil was still in good shape, might be perceived as being in poorer health; and an overgrazed range that had undergone extensive soil erosion would be an ecosystem in poor health. Obviously, judgments regarding "health" of ecosystems are a personal matter, and one person might regard an ecosystem as being in good health (if, for example, he was a farmer raising wheat in a particular field), while another might regard it as being in poor health (e.g., somebody who believed that the field should be allowed to revert to a wilderness state). Nevertheless, we do make value judgments about ecosystems, and society makes decisions about ecosystem management based on these value judgments.

Much of our present concern over the environment is because we can observe the health of many ecosystems undergoing a steady degeneration, even within the framework of the human time scale. Furthermore, many significant effects of human technologies on different ecosystems are related not to the technologies but to the ecosystems. Thus, when an agricultural technology that has proven highly successful in North America or Europe is introduced into central Africa or South America, it may prove disastrous because of the characteristics of ecosystems in these areas. The basic forces that govern natural ecosystems also govern human ecosystems, whether we like it or not.

Society depends on various ecosystems to produce food, fiber, wood, and recreation. Our dependence is sometimes greater than we like to admit. The very survival of human society, for example, requires that food-producing ecosystems yield enough food to feed the population well enough that they will not riot from hunger. Persistence of food-producing ecosystems is a foundation of modern society. The same could be said about quality of the drinking water supply, as well as other basic social needs derived from functioning ecosystems.

It is meaningful to distinguish two different kinds of human intervention into the environment. The first is deliberate management directed at meeting a perceived human need. Examples are agriculture, intensive forestry, fish culture, and park management. The second is the addition of pollutants that degrade the health of the ecosystem without increasing production of any socially useful product from it. Pollution is not related to managing the ecosystem; it is the waste product from a different managed system that is dumped into it. Air and water pollution are well-known examples.

The responses of ecosystems to management or pollution do not depend on whether they are natural. Indeed, the difference between a natural ecosystem and a human ecosystem is that the signals governing the structure and behavior of the natural ecosystem are natural in the first case and include some signals derived from society in the second (Figure 9.1). This is not to minimize the significance of people in ecosystems. But insects do not die any faster when a given concentration of nicotine is used as a pesticide than when it occurs naturally in the tobacco plant. Algal blooms occur just as readily in naturally nutrient-rich waters as in culturally eutrophic waters. Insect pests on cultivated crops destroy no more of the crop than they would if the same density of crop existed in nature. The fact is that pesticides are used in the human ecosystem, that water pollution does increase the nutri-

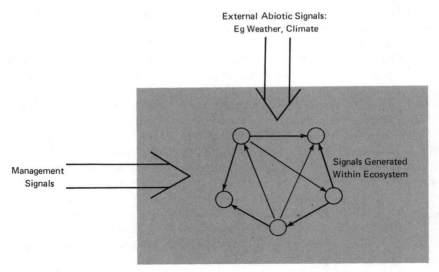

Figure 9.1
Schematic representation of the set of signals operating in an ecosystem. Circles represent ecosystem elements such as populations, soils, etc. They respond to all signals from all sources and generate the majority of signals found in the ecosystem. The external abiotic signals are the natural stresses found in all environments. Management signals are specifically input by human society.

ent content of lakes, and that crop monocultures have a much higher population density of prey plants than in any natural ecosystem. These comprise signals just as does any other toxic chemical in the ecosystem (of which there are many), any nutrient concentration produced by natural biogeochemical cycling, and any aggregation of individuals of a single species. The fact that people rather than nature generated the signals does not affect the way organisms respond to them.

What does make a difference is how the ecosystem responds to those signals and what their practical implications are. Specifically, what happens to the system when it is perturbed?

It is interesting to compare the responses of a complex constant community such as a tropical rainforest or a coral reef with those of a simple community such as the Arctic tundra. Both are buffeted by intense stresses: It is difficult to find any environmental force as powerful as the hurricanes and other storms that regularly affect coral reefs and at least some tropical and subtropical areas. In the same way, it is difficult to find any stress as intense as the great cold that pervades the Arctic winter. But the plants and animals that can survive in those ecosystems are adapted to the stresses naturally found there. The responses of the two ecosystems to stress are totally different. A complex community like the rainforest is so well buffered that it resists change. When pieces of the coral reef break, the reef can immediately begin to regrow as soon as the storm has passed. In either case, the observer would notice the results of the stress, but these would not appear to be very great. The same resistance that buffers the community from stress-induced change prevents it from ever reaching equilibrium. Parts of it are always dying as they are stressed and then regrowing, and stress comes too often to allow the community ever to reach a steady state.

The situation is different in the simple community that is not as well buffered. The ecosystem is not resistant to change, and relatively small perturbations can cause drastic changes in the ecosystem as a whole. But there is no long-term damage as long as the stress is within the limits of adaptation of the organisms constituting the community. Simple communities are resilient: They may change drastically, but they can also move quickly toward their equilibrium levels. Once again, it is doubtful that a simple community ever actually reaches its equilibrium and stays there. Stress is simply too omnipresent in the real world.

Ecological succession gives some useful insights into the responses of ecosystems to stress. Figure 6.12 points out how a successional sequence can be destroyed at any stage and taken back to the beginning, or at least to some "earlier" stage. How this occurs is not really important. It may be by fire, clearing of the area by people, or undermining of the ecosystem by flood or wind erosion. When the ecosystem is not being destroyed, however, it may change slowly or quickly, depending on the complex of signals to which the ecosystem responds.

In the textbook examples of succession on a sand dune or an old field, the community starts out simple and becomes more complex. It begins with a low degree of resistance and ends with a high degree. It is vulnerable and easily destroyed early on, but it recovers quickly. It is less vulnerable to destruction late in succession, but if destroyed must retrace the entire successional sequence. The external forces acting on the ecosystem do not care if it is simple or complex, young or mature. A bad storm is a bad storm, and it is simply more likely to destroy a youthful successional stage than a mature one.

The populations making up all ecosystems must be adapted to the range of variations found in their habitat. The species that constitute a rainforest or a coral reef or even a mature temperate forest are specialized to a narrow range of environmental variation. Niches are precisely drawn, and species can be very successful within them. Some latitude is necessary, of course, because some fluctuations are present even in the most predictable of environments. In the tundra or early successional temperate ecosystems, conditions are less predictable, and species must be able to withstand greater variation in order to survive.

The stresses that populations must routinely withstand are minor when compared with the perturbations introduced by people. These can easily change the gross environment of an area so that conditions are no longer within the range of tolerance of the organisms that formerly occupied the area. This is different from the destruction of an early successional ecosystem by a storm. In this case, the successional community can recolonize the area fairly quickly, and the natural process begins over. As pointed out eloquently by Joseph Connell and Ralph Slatyer, local destruction followed by conditions that allow recolonization must be quite common in nature.[2] The perturbations from human intervention may be of a different order of magnitude, so that normal recolonization is impossible. The alternative path that is actually followed may not be what the developer anticipates.

Let us take an example. Intensive agriculture has been markedly successful in many areas of the temperate zone. It was long thought that agricultural technologies developed in Europe and North America could be transplanted directly to the Third World tropics and subtropics to bring about a "green revolution" that would allow poor countries to feed themselves. The green revolution has had some successes. But its promises have also led to the clearing of vast tracts in rainforests such as the Amazon and parts of Asia. Many of the soils of these regions are a sort of oxisol called "laterite." When exposed directly to the sun, it can harden irreversibly into a surface more akin to concrete than a usable field. Even if *laterization* does not happen, clearing of nutrient-poor rainforests can disrupt nutrient cycling and lead to forest regeneration that is agonizingly slow.

At the other end of the spectrum, a major perturbation of a normally fragile ecosystem such as a tundra is equally damaging. When one clears vast areas of tundra, in order to mine minerals, build roads, or drill oil wells, the living community takes years and years to return. On Alaska's Seward Peninsula is a trail over which a single wagon ran twice in 1920. The vegetation still has not healed over those wagon tracks after 60 years. But failure of the tundra to reestablish itself after clearing is for a different reason from that of the rainforest. Because the environment is so marginal and unpredictable, organisms must be adapted to withstand great fluctuations, and it is unlikely that simple land clearance could alter the environment so much as to put it outside the tolerance ranges of the organisms in the community. More likely, the productivity of communities in such a harsh environment is so low that the rate at which they can recolonize an area is painfully slow.

Temperate zone ecosystems are intermediate in their predictability, and productivity. A major perturbation in the ecosystem can bring major changes in ecosystem composition, but the regional vegetation can reestablish itself relatively quickly once the perturbation has passed. Thus fields in temperate zones can be cleared for the planting of crops without destroying the soil, and normal succession can return it to natural vegetation fairly rapidly after it is abandoned. Analogous phenomena are noted in other productive areas characterized by wide fluctuations in the natural environment, such as estuaries. Perturbations of these ecosystems to a degree that would destroy a more uniform environment have relatively little effect. The resilience of variable-climate or temperate-zone ecosystems, then, is high, just as the resistance to change is moderate. These are the conditions under which artificial ecosystem structures based on perturbation by outside forces can be most persistent.

Figure 9.2 diagrams the variation of several environmental variables along a gradient from a rainforest to a tundra. Ecosystem resistance to change is highest where productivity, subdivision of niche space, and environmental constancy are highest, and it drops as any of these decline. Resilience is highest where productivity is relatively high and resistance is relatively low; it declines toward the tundra where productivity is low, and it declines toward the rainforest where the inertia from resistance increases. It is significant that the temperate zones are where ecosystem resilience tends to be highest, and resistance is also fairly high. It is little wonder that the most successful examples of human ecosystem management are found in these zones.[3]

How much stress can an ecosystem withstand before it collapses? There is no doubt that different levels of stress have different results in different ecosystems. For example, the mollisol grasslands of central North America are the most fertile and resilient croplands on the continent. They were cultivated for many years without fertilization, crop rotation, or other practices which were necessary on the alfisols and spodosols of the eastern forested regions. Yet overcultivation caught up with even the prairies, and their productivity began to decline in the early twentieth century, culminating in a massive reduction of ecosystem fitness during the Dust Bowl era.

It would be useful if there were some precise method to measure the health of an ecosystem and gauge its trends through time. To be sure, we can

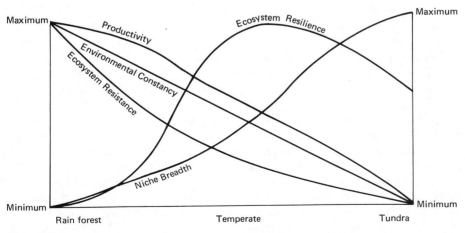

Figure 9.2
Schematic representation of variation in several ecosystem parameters along a broad environmental gradient.

quantify certain ecosystem characteristics and even optimize them for certain crops or other purposes. But we cannot yet quantify the aggregate notion that is an ecosystem, nor can we agree on which factors ought to be included. There are some working definitions of good or poor health for some specific ecosystems, but there is nothing that can be generalized throughout the world.

This lack is significant, since management has a key role in twentieth century society, and the future of many ecosystems depends on decisions made by people whose interpretations of ecosystem health include a heavy overlay of value judgments. Many of these people are scientists who have spent their lives studying ecosystems; others are politicians whose background is in the law or commerce. It is quite likely that none of them has all the answers. But they will (and must) still make decisions: What kind of aid should be given to a country in the wet tropics that wishes to improve is agriculture? What agricultural policy should a developed country have? How much virgin land is required to maintain the genetic basis of wild populations? Indeed, what is the value of these populations? What philosophy should be used to manage national parks, national forests, and other major environmental treasures?

There is no goal to natural ecosystems. Populations respond to a mix of signals from at least three separate networks. The primary network is the flow of energy and materials through the food web and the various biogeochemical cycles. The secondary network is information used by various animals to capture prey, consummate reproduction, and interact actively with other animals. The third is the gene pool of each population, which determines tolerances, instinctual responses, and so on. All of these networks show some feedback, but in varying time scales. Information can be interpreted quickly, but materials flow and gene frequency change are both very slow. The closest analogue to purpose in the system is natural selection: A population has a role in the community as long as it survives. When it loses out in the struggle for survival, it is gone. This is not to say that the *purpose* of the ecosystem is survival (as, for example, the purpose of a washing machine factory is to make washing machines). But it is entirely true that the *result* of the system's behavior is the survival of the populations it is composed of (just as the result of the behavior of the factory is a dock full of washing machines).

People do not need to wait for natural selection. We can measure the parameters of an ecosystem and decide whether it is healthy or not (by whatever criteria of health we wish to use). We can then make whatever management decisions we want and implement them almost before natural selection has had a chance to work. It should be absolutely clear to all observers that we have not bested natural selection. Our dependence as a biological species on ecosystems is so great that their collapse would spell disaster for human populations. People starve and social structures break when agricultural ecosystems collapse in a drought. One can see this clearly in looking back at the drought in the Sahel of Africa during the mid-1970s. Economies suffer when cash crops fail because of natural disasters. People get sick when pollution of water supplies supports disease organisms.

The laws that govern natural ecosystems are the laws that govern ecosystems dominated and managed by people. Valid management decisions should be based on these laws and recognize the characteristics of the ecosystem in question. Human technology is a powerful tool to influence the environment, but it cannot influence all of the factors to which living organisms respond. We cannot be sure that we are using technologies in the most appropriate way until we can predict the implications of their use before their use.

Despite our willingness to barricade ourselves behind centrally heated walls in sterile urban and suburban ghettos, we still depend on food produced through photosynthesis for our biological energy and mineral nutrients. We move through, reproduce in, and expel our waste products into the environment just like any other species on the face of the earth. As long as this is the case, we depend on ecosystems to provide these services to society. When the health of an ecosystem becomes so low that this is no longer possible, the populations inhabiting that ecosystem are in a precarious position.

Few of the ecosystems upon which we depend are natural. Except for wilderness whose justification is recreation, leisure, scientific study, or "knowing it's there," human ecosystems are engineered to serve some particular purpose, such as food production. They tend to be relatively simple, comprising a suite of populations that would not normally constitute an ecosystem. They tend to be unstable in that they would change their form if management were not able to generate a series of signals that counterbalanced the forces that led to change. In industrial societies, these management signals also depend on input of energy derived from fossil fuels or nuclear energy. Persistence can occur if the management signals are sufficient to maintain the ecosystem in a productive form.

It is tempting to visualize a managed ecosystem as not changing very much from year to year. This is, in fact, the way managers like to view the systems they control, whether these be factories, agricultural fields, or waste treatment plants. It is also unusual in nature. Natural ecosystems are in a constant state of flux as they respond to geographic variegation and climatic inconstancy, but the rate of change within the ecosystem is often slower than the rate of change of the outside signals. The result is a community that may appear to be constant. But its constancy reflects constant change in the influences governing the system rather than a balance of forces. It is not the constancy of a person sleeping in a hammock, whose weight is balanced by the supporting forces of the hammock; it is the constancy of the racquetball constantly being batted by players against the walls of the court and bouncing back into the zone of play.

Managing ecosystems effectively requires that we understand what signals society can generate, how signals from the regional climate, for example, affect the ecosystem, and how signals are generated within ecosystems (Figure 9.1). It is not reasonable to assume that management practices that have been successful in the past will always continue to be successful; nor is it reasonable to neglect signals that actually come from society simply because they are unintentional. For example, the classic example of the deliberately maintained ecosystem is the agricultural field. Essentially, one species is allowed to exist while other species are removed through expenditures of energy, herbicides, and pesticides, and the abiotic environment is controlled by extensive use of irrigation, fertilization, and tilling. The farmer attempts to block the natural successional forces that would normally return the ecosystem to its preexisting natural vegetation. Even so, the stable production of crops is by no means ensured, as indicated by periodic blights, which may destroy substantial portions of a crop.[4]

It is not so very different when alteration has been an unpremeditated by-product of other, intentional activities. Most unpremeditated perturbations are types of pollution, but there are others, such as weather modifications. Because these changes are unpremeditated does not imply that they are not important; indeed, they may have significant direct impact on people. By-products of exploitation have commonly affected ecosystems other than those being directly used. For example, agriculture leads to siltation in rivers, water pollution from fertilizer and pesticide runoff, and increased dust in the air. A city may draw its water from one part of a river and discharge its sewage into another, but the sewage discharge is almost always downstream from the city's water intake. The atmosphere and surface waters have been used for so long as the toilets of civilization that this use has become enshrined in the common folklore. LaMont Cole even reports a lawsuit in which one side contended that the herbicide 2,4-D should be considered a normal part of the Iowa atmosphere during the summer.[5] The result of these attitudes has been a degradation of air and water quality to such a degree that the health of many ecosystems, both terrestrial and aquatic, is greatly impaired.

One might be able to justify ignoring the degradation of ecosystem health if ecosystems of all sorts were not important to people. But they are. Not only do we depend on several different kinds of ecosystems for basic biological and economic needs, but all ecosystems are interconnected in ways that are not always obvious. DDT used on fields in the United States and western Europe got into the atmosphere and was carried throughout the world. Tests for DDT showed that no population was immune to DDT exposure—even Antarctic penguins.

In the same way, we do not know the significance of threats to oceanic photosynthesis from pesticides and other industrial chemicals. We do know that the oceans account for a very large portion of the world's photosynthetic oxygen production. We know that DDT and certain other pesticides reduce the rate at which oxygen is produced by photosynthesis in marine phytoplankton.[6] As it happens, the levels at which DDT depresses photosynthesis in phytoplankton are higher than its solubility in seawater; hence, DDT dissolved in seawater is unlikely to deplete the earth's oxygen supply. However, DDT is highly soluble in oil, and oil spills from tankers are becoming increasingly common. Whether or not DDT dissolved in oil spilled from tankers will pose a threat to the earth's oxygen supply is unknown. But we cannot write off the responses of marine phytoplankton to DDT and oil as being negligible.[7] Besides, there are other widely used pesticides whose effect in depressing the rate of photosynthesis in marine phytoplankton and whose solubility in water are both substantially higher than those of DDT.

Oxygen is not the only useful substance whose presence depends on ecosystems sometimes regarded as the proper dumping grounds of society. Many biogeochemical cycles span several major ecosystems, and perturbations that change the rates of import and export from one ecosystem to another can damage the health of both. A somewhat oversimplified example of this is given by the interaction of livestock feedlots and adjacent bodies of water. Livestock tend to be fattened in feedlots rather than on the range for economic reasons. On the range, the animals' urine and feces fertilize the land naturally to replace much of the nutrients removed by herbivory. However, the nutrients contained in the feed used in feedlots are removed from fields in their entirety, causing a reduction in soil fertility that must be made up with massive fertilization. Livestock excreta are too concentrated to be utilized by plants in the feedlot area and so are washed into nearby waterways where they become a pollution problem. Disruption of a normal biogeochemical cycle has led to the deficiency of materials in one ecosystem and a surplus in another, reducing the health of both. There are many similar instances.

One should not underestimate the role of values in decision making for ecosystem management. This shows up clearly in regard to wilderness or "undeveloped" land. Scarcely a month seems to go by that someone in a position of authority doesn't cast his eyes onto wilderness in mountain areas, tropical rainforests, or wetlands and see in them the solution to the world's energy or food production problems. Perhaps a politician can be forgiven for overlooking the fact that the reason that wilderness lands are still undeveloped is because they have historically been uneconomic to develop. If this were not the case, they would already be farmlands. But decisions of major import to the world, like opening up and clearing the Amazon, have been undertaken with a minimum of prior study and scientific understanding. The idea that they might fail is often not even considered.

Almost more important is that decision makers may not realize that some of the most significant values of ecosystems may not be the ones that seem clearest. For example, it is "obvious" that tropical rainforests represent a tremendous land area that should be cleared and turned over to raising cattle to supply hamburger to the fast food restaurants of North America. What is less obvious that a very large percentage of the world's pharmaceuticals come from living things that grow in tropical rainforests and will not grow on tropical cattle ranches.[8] "Development for the sake of development" is a value judgment that many countries (both developed and developing) have made. One must recognize the values on which the judgment was made (even if one disagrees with them). Judgments are based on values far more often than on scientific data or interpretations. But this does not make them perfect, or even good judgments for the society in question.

People are among the most widespread of creatures. We participate in more ecosystems than any other large animal, and we probably rival the most catholic of bacteria. There is no type of ecosystem in which we do not have at least some interest; changes in the health of any one of them affects us in some way. Pollution has not always been something we had to worry about. It used to be enough to dump wastes into a suitable ecosystem, and natural processes would take care of them for us. Indeed, the watchword of pollution control used to be "dilution is the solution to pollution."

Waterways, which have been dumping grounds as long as there have been materials in sufficient quantity to necessitate dumping, are also sources of fish and drinking water. At one time, the amount of refuse being poured into streams was small enough that detritivores in the streams could break it down within a fairly restricted area, with the remainder of the waterway being clean enough to provide fish and potable water. But the volume of wastes going into waterways has become so great that refuse decomposition through the detritus food chain has left large sections of major rivers so deficient in oxygen that they are essentially barren of fish. Nutrients introduced into waterways through sewage have caused extensive eutrophication in systems as large as the Great Lakes of North America. Indeed, large portions of the hypolimnion of Lake Erie become anoxic every summer, and there is intense concern over the long-term health of Lake Michigan.

Likewise, there was a time when most wastes introduced into the atmosphere were quickly diluted to concentrations low enough that they caused little damage. But airborne wastes are now so extensive that air pollution damage is widespread. For example, smog derived from automobile exhausts in the Los Angeles area is killing ponderosa and Jeffrey pine trees in the San Bernardino National Forest some 60 miles to the east, and it has eliminated spinach as an agricultural crop that can be grown in the Los Angeles area. Recreation, an essential of modern society, is dependent in large part on open areas, yet available open areas are being destroyed at an increasing rate. Sulfates produced by burning coal in industrial areas in the Ohio River valley and in Great Britain are being carried hundreds or even thousands of kilometers by the winds and transformed to acid rain that damages lakes and streams in the Adirondack Mountains of New York, in the Canadian Shield, and in Scandinavia.[9]

Modern society has become much more complex than it used to be, and the volume of wastes being produced is simply too great to be accommodated by the assimilation capacity of real-world ecosystems. Even if we were willing to accept a certain amount of degradation of the health of some areas to accommodate pollution, we cannot ensure the security of other ecosystems in which our best interest is to keep them clean.

Ecosystems and the Future

Most people recognize that the environmental crisis is significant. But it is hard to gauge just how important it is. People used to try to estimate the potential longevity of our civilization. The numbers varied from a pessimistic 10 to 35 years to an optimistic forever. Actually, this kind of crystal-ball analysis isn't as important as trying to recognize how to combine the scientific knowledge we do have with the values that govern decision making to yield a set of management strategies for ecosystems that will serve for high-density industrial civilizations.

We have made a lot of progress in the last 10 to 15 years. The area of Lake Erie that is anoxic in the summer is smaller now than it used to be. It is safe to swim in the Potomac River at Washington, D.C. Lake Washington in Seattle has had so much sewage diverted from it that it has improved markedly. But the rate of species extinction is still at worldwide epidemic levels, and many people still speak as though environmental principles had no relevance to decisions regarding land management.

Nobody knows what the future will hold for us. We have seen formerly healthy large-scale ecosystems such as Lake Erie, the Dust Bowl, and the estuaries of a dozen North American and European rivers deteriorate. It is both easy and justified to become cynical, and it is clearly naive to feel optimistic. Fish kills in the Mississippi or the Rhine or massive pesticide poisoning of robins, eagles, falcons, and pelicans are different from famine or plague in a city. But the materials that cause fish-kills and bird-deaths are consumed by people as well as by fishes and birds. The significance of the symptoms of environmental degradation go far beyond the aesthetic.

Society has recognized that natural ecosystems have values that are not always easy to put into quantitative or economic terms. For reasons that people cannot always explain, a clean and safe environment has entered the value structure of industrial societies, so that decisions can be made to protect the environment on a value basis even if we cannot always point to obvious economic benefits. This is fortunate, but it does not lessen the need to understand the scientific principles governing ecosystems and to understand how society can manage stable human ecosystems by generating signals that are compatible with those produced in the ecosystem.

Notes

[1] Harrison, G. W., 1979. Stability under environmental stress: Resistance, resilience, persistence, and variability. *Am. Nat.*, **113**, 659–669.

[2] Connell, J. H., and Slatyer, R. O., 1977. Mechanisms of succession in natural communities and their role in community stability and organization. *Am. Nat.*, **111**, 1119–1144.

[3] Clapham, W. B., Jr., 1976. An approach to quantifying the exploitability of human ecosystems. *Human Ecology*, **4**, 1–30.

[4] Tatum, L. A., 1971. The southern corn leaf blight epidemic. *Science*, **171**, 1113–1116; Horsfall, J. G., et al., 1972. *Genetic Vulnerability of Major Crops*. Washington, D.C.: National Academy of Sciences.

[5] Cole, L. C., 1970. Playing Russian roulette with biogeochemical cycles, in Helfrich, H. W., ed., *The Environmental Crisis: Man's Struggle to Live with Himself*. New Haven, Conn.: Yale University Press, 1–14.

[6] Wurster, C. F., Jr., 1968. DDT reduces photosynthesis by marine phytoplankton. *Science*, **159**, 1474–1475; Menzel, D. W., Anderson, J., and Randke, A., 1970. Marine phytoplankton vary in their response to chlorinated hydrocarbons. *Science*, **167**, 1724–1726.

[7] Singer, S. F., 1970. Will the world come to a horrible end? *Science*, **170**, 125.

[8] Myers, N., 1979. *The Sinking Ark: A New Look at the Problem of Disappearing Species*. New York: Pergamon Press.

[9] Likens, G. E., Wright, R. F., Galloway, J. N., and Butler, T. J., 1979. Acid rain. *Sci. Am.*, **241(4)**, 43–51.

Epilogue

There is an old technique used in the coal mines. The miners take a caged canary into the mine because canaries are much more sensitive to toxic mine gases than people. When the canary dies, it is time for the miners to leave. The analog between the coal mine and the biosphere is clear: We live in a tiny portion of the total solar system, and we share it with many other organisms spread through many interconnected ecosystems. We know much about how natural ecosystems operate and how they are interconnected. We also know that the health of many ecosystems is deteriorating at an alarming rate. The symptoms of this deterioration—the fish-kills, the bird-deaths, the increasing algal blooms, the replacement of high-quality fish by coarse fish—are all analogs to the death of the canaries in the coal mines. But the miner can get out of the mine. People cannot leave the biosphere.

Further Reading

Cox, J. L., 1970. DDT residues in marine phytoplankton: Increase from 1955 to 1969. *Science*, **170**, 71–72.

Dasmann, R. F., 1968. *Environmental Conservation*, 2nd ed. New York: John Wiley & Sons, Inc.

Ehrenfeld, D. W., 1970. *Biological Conservation*. New York: Holt, Rinehart and Winston.

Ehrlich, P., and Ehrlich, A., 1981. *Extinction: The Causes and Consequences of the Disappearance of Species*. New York: Random House.

Farnesworth, E. G., and Golley, F. B., 1974. *Fragile Ecosystems: Evaluation of Research and Applications in the Neotropics*. New York: Springer-Verlag.

Frankel, O. H., and Soule, M. E., 1981. *Conservation and Evolution*. New York: Cambridge University Press.

Margalef, R., 1968. *Perspectives in Ecological Theory*. Chicago: University of Chicago Press.

Meadows, D. H., Meadows, D. L., Randers, J., and Behrens, W. W. III, 1972. *The Limits to Growth: A Report for the Club of Rome's Project on the Predicament of Mankind*. New York: Universe Books.

Meggers, B. J., 1954. Environmental limitation to the development of culture. *Am. Anthrop.*, **56**, 801–824.

Pomeroy, L. R., 1970. The strategy of mineral cycling. *Ann. Rev. Ecol. Syst.*, **1**, 171–190.

Rosenzweig, M. L., 1971. Paradox of enrichment: destabilization of exploitation ecosystems in ecological time. *Science*, **171**, 385–387.

Soule, M. E., and Wilcox, B. A., 1980. *Conservation Biology: An Evolutionary–Ecological Perspective*. Sunderland, Mass.: Sinauer Associates.

Woodwell, G. M., Wurster, C. F., Jr., and Isaacson, P. A., 1967. DDT residues in an east coast estuary: A case of biological concentration of a persistent insecticide. *Science*, **156**, 821–824.

Index

Intrinsic rate of natural increase, 111
Ions, 6
Isle Royale, Michigan, 94

J

Japan, 73

L

Lady's slipper, 140
Lake Baikal, 198
Lake Erie, 82, 158, 272
Lake Mendota, Wisconsin, 27, 35,
  36, 195
Lake Superior, 64
Lakes, 184–198
Landforms, 217–218
Laterite, 236, 266
Leaching, 225, 232–233
Legumes, 58
Lepomis macrochirus, 49
Lichens, 88
Life cycle, Life histories, 49, 112–114
Life zones, 188–191, 202–203
Light, 23–25, 172–173
Light compensation level, 173
Limestone, 54,
Limiting factors, 50–52
Littoral zone, 190–191
Little Sioux River, Minnesota and
  Iowa, 87
Living organisms, 10–14
Logistic curve, 111
Luxury uptake of nutrients, 176

M

Macronutrients, 50
Massachusetts, 141
Mesotrophic lakes, 193–198
Metalimnion, 186–188
Mice, 103
Microhabitat, 135, 239
Micronutrients, 50
Midoceanic ridges, 203
Mites, 92, 114–115, 160

Mollisols, 235, 237
Mollusks, 37
Monarch butterflies, 96
Moose, 94
Mosquitoes, 117
Mountain lions, 79, 99
Mutualism, 88–90
Mycorrhizae, 90
Mytilus californicanus, 48
Myxoma virus, Myxomatosis, 117

N

Natality, 122
Natural selection, 10–14, 79
Net primary production, 30
Neutralism, 83–84
New Zealand, 51–52
Niche overlap, 136–138
Nitrification, 57
Nitrobacter, 57
Nitrogen, 174–175
Nitrogen cycle, 56–58
Nitrosomonas, 57
Nonequilibrium systems, 142, 155
Nucleation particles, 67
Nutrients, 6, 7, 46–64, 71, 85,
  199–200, 232

O

Oceans, 31, 71, 143, 171, 201–210
Oceanic circulation, 71–72
Oligotrophic lakes, 193–198
Olson, Jerry S., 153
Oneida Lake, New York, 196
Ontario, 143
Opportunist species, 112, 125
Optimum, 47, 132
Opuntia, 93–95
Orconectes species, 87
Oryctolagus cuniculus, 116–117
Overturn, 186–188
Owls, 144
Oxisols, 235–236
Oxygen,
  8–10, 25, 27, 41, 174, 208, 211
Ozone, 8–10, 25